Trauma-Informed, Culturally Based Intervention

TICBI

Trauma-Informed Culturally Based Intervention

Nimo Bokore

Trauma-Informed, Culturally Based Intervention

Integration of Neuroscience and Social Work in Supporting Refugees

Springer

Nimo Bokore
School of Social Work
Carleton University
Ottawa, ON, Canada

ISBN 978-3-032-06834-7 ISBN 978-3-032-06835-4 (eBook)
https://doi.org/10.1007/978-3-032-06835-4

SSHRC Partnership Engage Grants

This Springer imprint is published by the registered company Springer Nature Switzerland AG
The registered company address is: Gewerbestrasse 11, 6330 Cham, Switzerland

To My Parents
Fadumo Gatur and Sheikh Ali Bokore: You are my two role models in life.

To My Children
Ahmed, Meron, Aragson, and Awo-Ann: You have taught me what resilience looks like and have always motivated me to overcome difficult circumstances and do more.

To My Grandchildren
Jasmine Wehelson, Hanna, Zoe, and Jibi (Zak): You have taught me that discontinuing intergenerational trauma and fostering lasting resiliency and strength is possible.

Preface

Addressing the trauma intervention needs of our multicultural society necessitates a cultural understanding of the diverse experiences and trauma responses. Over the past decade, there has been an increase in writing about trauma-informed or culturally based interventions. However, no single book comprehensively explains these interventions in social work practice. This book aims to fill that gap at the micro level, serving as a resource for training graduate students as they examine future work with survivors. It provides a step-by-step intervention process that begins with understanding the survivors' histories and the impact of their trauma, and includes healing activities. Anonymous case studies of war survivors illustrate how to conduct trauma-informed culturally based intervention (TICBI). Each chapter provides a detailed guide for practitioners using TICBI, accompanied by interactive activities that allow practitioners to practice the skills learned in each chapter.

At the macro level, this book contributes to knowledge by identifying sustainable solutions for human displacement. In this context, sustainable solutions refer to a long-term strategy centered on helping survivors recover and heal, enabling them to live safely and with dignity. It provides intervention tools that empower survivors to take control of their lives, regardless of their circumstances. This approach facilitates individual intervention methods that support survivors, whether they reside in refugee camps, are new to a country, are working on their economic and social integration, or are assisting those wishing to participate in voluntary repatriation programs to return and rebuild their lives in their home countries.

In recent years, "trauma-informed" and "culturally based" have become buzzwords used in various disciplines, without an in-depth explanation of what they entail. In this social workbook, as a survivor of war and traumatic experiences with extensive training in trauma intervention, I recognize the challenges of using these words without substance. This book aims to move students from a superficial use of these terms to a deeper understanding of trauma, culturally based meaning-making processes, and a genuine commitment to improving the well-being of survivors. The intervention approaches in this book are designed to take students and practitioners beyond buzzwords by providing a step-by-step, proactive, collaborative, and evidence-based intervention.

Evidence-based trauma interventions in social work offer multiple benefits for survivors, enhancing the service provider's knowledge of trauma, the healing process, and improving practice outcomes. For service providers, understanding the neurobiology of trauma clarifies the changes occurring in the brain and body due to traumatic experiences. This knowledge allows them to plan and conduct interventions, mitigating the long-lasting effects of trauma, which can lead to mental and physical health challenges. The insights into the neurobiological aspects of trauma impacts and the healing process discussed in this book also aim to inform and inspire social workers who hesitate to engage with neuroscience. This book can expand their professional scientific knowledge, thus enhancing their accountability in trauma interventions and enabling informed, evidence-based decision-making. Social workers must learn and apply the latest research findings to improve their clinical expertise, mitigate work-related stress, and deliver practical and tailored services to survivors.

Evidence-based knowledge for clinical social work benefits from the latest advances made in neuroscience. These advances can be integrated into social work's graduate programs to enhance the teaching of mental health and trauma intervention courses. Understanding the depths of trauma and cultural interpretation is crucial for planning interventions focusing on how people think, feel, learn, change, and heal from the adverse effects of trauma. Although integrating social work practice and neuroscience is vital to clinical intervention, few social work authors have published on the subject. These publishers, without extending what the link looks like in practice, each explain the benefits of this knowledge to the field. For example, Jeffrey Applegate and Janet Shapiro (2005) point out how social work has recently embraced biopsychosocial assessment in clinical interventions and how the biological aspect, or the knowledge base for practice, remains underdeveloped. Acknowledging the benefits of neuroscience to social work, Rosemary Farmer (2008) identifies neuroscience as the "missing link" in biopsychosocial assessment and intervention. Similarly, Deb Dana (2021) links the benefits of understanding polyvagal theory and autonomic responses in trauma intervention. Following these few social work authors, I use my personal experiences of trauma, practice in the field, and research to develop a method for linking social work and neuroscience when providing the trauma-informed, culturally based intervention (TICBI).

As a survivor of war and multiple displacements and forced migration experiences, I view trauma responses and the necessary interventions differently. I recognize how my cultural learning shapes my feelings, reactions, and interpretations of life challenges, including traumatic experiences. While examining human displacement and resettlement challenges for my PhD study, I discovered the varied responses of survivors that influence their reactions to past or present distress. Over the years, while conducting various research projects, I came to understand how frequently survivors express their experiences and feelings based on their cultural and religious values. In clinical intervention, it is also essential to understand the neurobiological aspects of trauma, including its impact on the body, brain, and mind. I categorize survivors' responses into three categories, which I call mindsets: "survival-mindset, therapeutic-mindset, and active, engaged mindset," to define the varied responses and trauma-informed, culturally based intervention (TICBI) designed to assist survivors in their recovery. To explain this process, this book is structured into five sections, each containing two chapters.

The first section focuses on the global refugee crisis and the stages of forced migration. In Chap. 1, I explore the impact of the ever-expanding global wars, immigration policies, and public perceptions affecting the human rights response to displacement and resettlement. In Chap. 2, I discuss the three stages of forced migration, using the stories of research participants to illustrate the emotional and physical impacts of these experiences. In Stage one, or the pre-migration stage, I examine the individual's response to their experiences and the effects those experiences have on their body, brain, and mind. In stage two of forced migration, also known as the transit or devastating stage, I discuss the emotional and physical challenges related to forced displacement. In the third and final stage of forced migration, which I refer to as the resettlement stage, I examine the challenges of resettlement, their impacts, and various responses to these challenges.

The second section focuses on the neurology of trauma, discussing the changes in the body and brain that occur in survivors, as well as the risks associated with trauma, transference and intergenerational trauma, and the development of post-traumatic stress disorder (PTSD) or post-traumatic growth (PTG). In Chap. 3 of this section, I begin exploring the negative impact of traumatic experiences, including how survivors perceive threats, regulate emotions, and form memories that affect their well-being and contribute to developing post-traumatic stress disorder (PTSD) and the healing process. In Chap. 4, I highlight the positive aspects of traumatic memories, including the emergence of post-traumatic growth (PTG). This enables survivors to adapt by transforming their views of memories, fostering their current personal growth, strength, and resilience.

In the third section, I address the challenges of trauma care, including the current service gaps due to funding cuts, increasing service demands, and the impacts of cultural barriers in trauma intervention. In Chap. 5, based on my study, the TICBI study, and service providers' responses, I discuss job demands, including the experiences of burnout and vicarious trauma. In Chap. 6, I discuss the role of culture in trauma response and the importance of acknowledging the role it plays in trauma intervention.

In the fourth section, I explain the pre-planning and initial assessment phases of TICBI. As a result, in Chap. 7, I discuss the importance of planning TICBI, focusing on creating a safe, supportive, collaborative, and culturally informed environment. This planning involves utilizing knowledge and information about the client to provide services tailored to the survivor's history, experiences, and individual needs, thereby offering targeted interventions to meet each survivor's unique needs. In Chap. 8, I cover the initial evaluation of service needs, which includes an in-depth biopsychosocial and structural assessment that considers the biological, psychological, social, and structural factors affecting the survivor's well-being. Although the biopsychosocial assessment is championed in social work, the role of systemic structures in experiencing trauma is often overlooked in clinical intervention. This chapter discusses the importance of evaluating the role that underlying structures have in triggering or maintaining the survivors' traumatic responses.

In the fifth section, I explain the change phase of TICBI by providing a step-by-step intervention, along with an additional tool that can be utilized at each step. In Chap. 9, I discuss various intervention methods, based on individual needs, to promote recovery using case studies. Each intervention is tailored to the survivor's specific culture and situation, recognizing that these methods may not apply to all cultures but can be adapted to individual needs using the tools provided in Chap. 10.

I focus the whole book on TICBI research, which is beneficial for working with ethnic minority refugee groups resettled in Western nations, particularly Canada. The book will provide valuable TICBI knowledge and examples, enabling this evidence-based textbook to support those working in practice settings and others interested in trauma-informed, culturally based research on clinical interventions.

The book is intended for graduate students as a textbook in clinical intervention. It will be a valuable resource for training graduate students interested in working with diverse groups, such as Indigenous peoples and other minority or marginalized communities living in refugee camps or resettled in Canada, the USA, the UK, Australia, and New Zealand. The book will also benefit researchers conducting studies on refugees and trauma intervention across cultures.

Realizing the need for a new book linking social work and neuroscience in trauma intervention, I devote the entire book to understanding the needs of survivors who often remain vulnerable and marginalized. This book addresses that need, focusing on the knowledge and skills necessary to work with these groups to help them recover from their past ordeals and present challenges.

Ottawa, ON, Canada — Nimo Bokore

Acknowledgments

Like any publication, this book would not have been possible without the support of the co-investigators who participated in the early data collection and analysis, as well as the research assistants who collaborated with me throughout various stages of the TICBI project. I also want to express my gratitude to my editor and the team at Springer, who believed in the value of this book. Thank you all. Additionally, I would like to sincerely thank the Social Sciences and Humanities Research Council of Canada (SSHRC) for awarding me the Individual Partnership Engage Grants that funded the preliminary research project. Thank you all.

Competing Interests The author has no competing interests to declare that are relevant to the content of this manuscript.

About the Author

Nimo Bokore, PhD, is an Associate Professor at the School of Social Work at Carleton University in Ottawa, Ontario, Canada. She previously taught at York University in Toronto, Ontario, and at Toronto Metropolitan University, formerly known as Ryerson University, before joining Carleton University. Nimo has worked with community organizations and health centers that serve immigrants and refugees. Over the past fifteen years, she has taught undergraduate and graduate students, conducted both local and international research focused on refugees. She is a member of Carleton's "Local Engagement Refugee Research Network (LERRN)," which promotes refugee protection and durable solutions. In the past six years, as the lead for student training at LERRN, she has taught and mentored graduate, postgraduate, and early-career researchers with a focus on cross-cultural and cross-border research methods.

In her research, Nimo, a survivor of war, concentrates on issues related to refugees, forced displacement, trauma, mental health, trauma transference, post-traumatic stress disorder (PTSD), and post-traumatic growth (PTG). She has led and co-led several research projects and published multiple papers, as well as two book chapters on war, trauma, and refugee resettlement: "Understanding the Refugee Experience in the Canadian Context." In: Sethi, B., Guruge, S., Csiernik, R. (eds.) *Syrian Refugee Reflections on Canada's Sponsorship Programs*. Palgrave, and "Forced Migration, Memory and Testimony." In: Meerzon, Y., Dean, D., McNeil, D. (Eds.) *Migration and Stereotypes in Performance and Culture*. Cham, Switzerland: Palgrave Macmillan.

List of Figures

List of Tables

Part I
The Global Refugee Crisis and Stages of Migration

Chapter 1
The Global Refugee Crisis

1 The Global Refugee Crisis

Human migration, whether forced or undertaken for economic reasons, is a complex phenomenon with a long history of debate among host countries about safety, race, and demographic replacement. This debate continues to ignite nationalist ideologies based on protecting ethnic and racial identities. It also works to elide the difference between refugees, who leave their homes under the threat of death and need humanitarian assistance, and economic migrants, who, in the push-and-pull factor between migrant-sending countries and host or receiving countries, leave behind unfavorable living conditions at home and seek opportunities in other, often developed, countries.

As nationalistic debates continue in host countries, refugees forced to leave their homes live in precarious conditions as undocumented and stateless residents of countries other than their own or in refugee camps, which are scattered around the world. What is missing in these protectionist debates is an understanding of the trauma and emotional crisis refugees go through in their attempts to cross borders for safety and an acknowledgment that their lives are in danger. From 1977 to 1978, the Ethiopian-Somali War was fought in East Africa over the Eastern Province, also known as Region Five. The war destabilized the region, displacing many Ethiopian-Somalis in the late 1970s and early 1980s. Over 2.97 million Somalis from the area were internally displaced, and 650,000 remained in camps in Kenya, Uganda, and other countries for decades (Ager, 1999; Kagwanja, 2000; Milner, 2009).

The Middle East is home to the oldest and most protracted refugee crisis—the Palestinians, who were expelled 70 years ago from their homes, and now, multi-generational groups live in numerous camps in Palestine, Gaza, and neighboring Arab countries (Ager, 1999, p. 114). The pro-democracy protests known as the Arab Spring (2010–2011), challenging authoritarian regimes in Arab nations in the Middle East and North Africa, created another refugee crisis. More than 6.8 million

N. Bokore, *Trauma-Informed, Culturally Based Intervention*,
https://doi.org/10.1007/978-3-032-06835-4_1

Syrians have been displaced, and the same number of people remains internally displaced. Before the Arab Spring, the US invasion of Iraq displaced thousands of Iraqis from their homeland. Some fled to neighboring countries, and nearly 1.2 million were internally displaced (UNHCR, 2022).

The largest refugee crisis in Asia is the displacement of Rohingya groups, which the UN describes as the most marginalized and victimized minority in the world. Thousands have fled their homes in Myanmar's Rakhine State to escape armed attacks, extensive violence, and serious human rights violations. There are now over 900,000 Rohingyas living in the Cox Bazar region of Bangladesh. They walked for days through dangerous jungles or crossed the Bay of Bengal in unsafe boats, facing death at each step to reach Bangladesh. While Rohingyas are the most recent group of displaced people in Asia, the ongoing crisis in Afghanistan worsened significantly when the United States withdrew its army from the country in 2022, increasing the number of displaced Afghans to 3.5 million (UNHCR, 2022).

In Latin America, a recent refugee crisis involves Venezuela, which has hosted refugees in the past. Still, it is now one of the largest refugee-producing countries on the continent. Government sources report that the number of displaced Venezuelans in the past few years alone surpassed 7.8 million globally, with the vast majority of them hosted in Latin America and the Caribbean countries (UNHCR, 2023). With this, Venezuela has become the second-largest refugee-producing country in the world in recent years (UNHCR, 2023).

In all these locations, many children have been separated from their parents for a prolonged time and remain refugees for years. While in refugee camps, these children grow up without proper nutrition or emotional support and face daily violence, creating intergenerational suffering. The UNHCR (2021, 2024) estimates that in 2018–2023, the number of children born in refugee camps reached 339,000. Most of those in Global South countries will likely remain refugees their entire lives.

The refugee crisis, however, is not limited to the Global South, as seen in the flight of Ukrainians in 2022 and Bosnian and Kosovar refugees in the 1990s. These refugees also highlight the differential treatment given to refugees from the Global North. When Syrian mothers were trying to cross borders to Europe, they faced barbed wire and violence. In contrast, Ukrainian mothers were welcomed by Europeans with baby strollers, food, shelter, and other basic needs. In mid-2022, when more than 7.3 million Ukrainians crossed borders into neighboring countries, there were no public debates about putting up fences or increasing border control, nor was there public discourse about the danger of accepting refugees.

A similar reaction was seen in the 1990s when asylum seekers from Bosnia and Kosovo began to cross borders. European Union members increased their humanitarian response by modifying their border restrictions and barriers to entry. Implementing the temporary protection laws allowed under the 1951 UN Refugee Convention decreased the level of trauma for Bosnian and Kosovar refugees. It provided opportunities to rebuild their lives in settlement countries. While durable solutions like this are extended to European refugees, restrictive humanitarian responses and coercion to "voluntarily" return are enforced for others, violating Europe's humanitarian obligations under the 1951 UN Refugee Convention

(Dahlman, 2016). This fear-based exclusion of certain refugees has a long history in Europe. Its reach extends beyond preserving cultural and religious values or racial identities, for example, the relocation of populations between Turkey and Greece based on ethnicity, the atrocities targeting Jews during the Second World War, and the failure of countries, Canada included, to accept Jewish refugees (Goździak et al., 2020). This history helps us understand how the humanitarian response to the refugee crisis is often based on political ideology.

Another example is Syrian refugees crossing the Mediterranean in unsafe boats. Initially, this group of refugees was invisible, unlike any other Global South refugees. The stories of their perilous journeys were ignored until the Syrian refugee child Alan Kurdi washed ashore on September 2, 2015. Seeing the image in the media of a dead child who resembled a native European child humanized the plight of refugees and softened public views. It opened doors for Syrian refugees to settle in Western countries, including Canada.

Selective humanitarian responses, degrees of silence, and apprehension about the arrival of those seen as strangers have existed throughout human history, often concealed in us-versus-them rhetoric despite ongoing promises of humanitarian assistance (Paynter, 2022). As Said (1979, 1993) wrote, labels of inclusion and exclusion are justified by devaluing those seen as Other, non-European, and non-Christian, overriding the needs of refugees fearing for their lives.

2 Refugee Resettlement in Canada

Canada's acceptance of those seen as Other aligns with the historical policies and public perspectives of Western countries regarding the limited welcome of all migrants and refugees. As shown in Table 1.1, throughout various periods in Canadian history, restrictive immigration policies targeted specific groups to reduce the number of non-White or non-European settlers in Canada. One notable restrictive policy was the 1885 Head Tax imposed on Chinese immigrants (The Canadian Encyclopedia, 2025). Changes to immigration policy in the twentieth and twenty-first centuries have permitted many non-European immigrants and refugees to resettle in Canada. The following chart illustrates the history of refugee resettlement in Canada.

2.1 *Exception and Exclusion*

Canada's migrant and refugee resettlement criteria are rooted in race-based, nation-building, benefiting White settler society. These preferences, or selections, still impact those seeking humanitarian assistance. Although Canada continues with its UN commitment and obligation to resettle refugees, there are visible inclusion and exclusion criteria in its policies, border restrictions, and reception of newcomers.

Table 1.1 Canada's history of immigration

Quakers 1770–1779	Pacifist Christians, and advocates of social justice and international relief, were persecuted for their beliefs in America. Quakers refused to take up arms during the American Revolution in the 1770s, and they were resettled as refugees in Ontario and Quebec
Black Loyalists 1780–1789	Blacks escaping slavery in the United States and Black Loyalists, most of them supporting the British cause during the American Revolution, came as refugees with resettlement promises of "freedom and a farm"
Jews 1870–1899	Thousands of European Jews escaping pogroms and social and economic challenges brought on by industrialization arrived in Canada as refugees
Ukrainians 1900–1939 1940–1949 2022–2023	Thousands of Ukrainians came to Canada in 1919 as refugees. Through lobbying efforts of this early group, 35,000–40,000 Ukrainians were settled in Canada after WWII (1945–1952). Since the latest Russian invasion of Ukraine, about 166,000 additional Ukrainians have arrived in Canada through the Ukraine Authorization for Emergency Travel visa.
Indochinese 1979–1980	Following the fall of Saigon in 1975, about 2.5 million Indochinese civilians, referred to as *boat people*, resettled in Western countries (mainly North America and Europe). Due to Canada's 1970s immigration changes, more than 60,000 refugees from Cambodia, Laos, and Vietnam were resettled in Canada
Kosovars 1999–2001	The Kosovar war between Serbs and ethnic Albanians displaced a large number of Kosovars. In 1999, about 7000 Kosovar refugees were airlifted to military bases in eastern Canada under the code name Operation Parasol. Two-thirds of those airlifted opted for permanent resettlement options and moved to cities across Canada
Bhutanese 2007–2015	Canada is one of the Western countries that made a multi-year commitment to resettle Bhutanese refugees of Nepalese descent living in refugee camps in Nepal. Since 2015, Canada has resettled about 6800 Bhutanese refugees
Iraqis 2009–2015	The US invasion of Iraq displaced more than 2 million people and killed about 600,000. Canada pledged to resettle 23,000; by 2015, it had resettled more than 25,475
Syrians 2015–2016	In 2015–2016, millions of Syrians fled their country seeking refuge in neighboring countries, primarily Lebanon, Jordan, and Turkey, and more than 4 million Syrians were registered as refugees by the UNHCR. During this time, Canada resettled more than 100,000 Syrian refugees

Adapted from these sources: Epp, M. (2017), Government of Canada-IRCC report (2022, 2024), Government of Canada, Statistics Canada (2024b), The Canadian Encyclopedia (2025)

Othering systems have been used throughout Canada's history, whether of Indigenous groups, former enslaved Blacks, or non-European groups. There are inclusion and exclusion criteria in the Immigration Act of 1906, Canada's first immigration policy, which gave the federal government power over the nation's borders. The act addressed nationalistic concerns, such as the protection of the ethnic composition of the country, by strengthening exclusionary policies restricting new arrivals from non-European countries. Similar trends were present in the United States, most notably, the 1908 edition of the "continuous journey regulation," designed to curtail Japanese and South Asians from entering the country. Canada's 1910 Immigration Act, an amendment to the 1908 regulation, prohibited those without a ticket or travel document from their native land from entering the country (Isajiw, 1999).

Canadian immigration policies are similar to those of many other Western nations. They establish the framework and mechanisms to address identity, ethnic representation, and integration. Each government sets the tone for how it addresses the structure of the population and ethnic identity through policies and programs (Whitaker, 1991). Exceptions and exclusions in Canadian immigration policies and programs continue to significantly impact who can resettle in Canada under its humanitarian program. A notable example is Canada's response to the Syrian refugee crisis, which involved resettling 101,045 refugees from 2015 to 2024, a commendable humanitarian action consistent with the 1951 UN declaration (Government of Canada, Statistics Canada, 2022). This response did not result in other racialized refugees being resettled, nor did it alter immigration policies that implicitly persist in Canada, determining who is accepted and who is excluded, leaving certain groups in prolonged refugee situations for generations. There is an urgent need to find lasting solutions to the human cost of displacement and suffering (Ager, 1999; Milner, 2009; Papadopoulos 2002, 2006, 2021).

2.2 Canada's Continued Humanitarian Commitment

A critical shift in Canada's refugee response was its endorsement of the 1951 UN Refugee Convention. In 1967, the then Minister of Manpower and Immigration, Allan MacEachern, presented to the House a protocol aligning Canada's interests with refugee resettlement and international law (Milner, 2009). The policy was finalized in 1970, introducing new immigration regulations, including the point system for resettling immigrants. However, refugee resettlement still lags behind (Milner & Wojnarowicz, 2017).

Policy changes have been more political than humanitarian throughout Canada's refugee resettlement history. A recent example is the Canada-Ukraine Authorization for Emergency Travel program, implemented by the Liberal government one month after Russia invaded Ukraine in February 2022, to fast-track immigration for Ukrainians. This visa was a modified version of the Temporary Foreign Workers Program (TFWP), which grants a six-month stay in the country. In contrast, Ukrainian refugees were allowed to remain in Canada for three years. Additionally, after three years, a Ukrainian can apply for permanent residency and eventually citizenship, a pathway that temporary foreign workers do not have. For Ukrainians, public and government support further eases their transition into permanent resettlement. The only visible challenge for this group is that they are eligible for government support only once their permanent residency application has been approved. This means that, until then, they rely solely on employment income, family funds, and other sources.

The preferential treatment of certain groups in Canada's humanitarian response to the refugee crisis, which is based on race and geopolitics, is in line with other

Western countries. For example, historically, Canadian immigration was based on a soft version of the nineteenth-century American Jim Crow model to deter escaping enslaved Black Americans from resettling in Canada. This group of Blacks was denied entry into Canada because they were considered ill-equipped to live in the Canadian climate. This rationale was often used, along with criminality and cultural oddity, to justify race-based and demographic selection (Mathieu, 2010, p. 6). When Blacks first arrived in Canada, policymakers stoked fear in the Canadian public through the media, perpetuating the belief that Canadian sovereignty and white settler identity needed to be protected. The few Blacks allowed to settle in Canada faced economic and social isolation. Their basic needs, such as land, which white settlers were entitled to, were denied to them, and they were paid meager wages to work for white settlers (Whitfield, 2006).

Current public sentiment about immigration in Canada parallels its history, and debates over migration and border security continue to instill fear of the Other. This view contradicts the humanitarian spirit of the 1951 UN Refugee Convention. Unfortunately, while these debates continue, so do restrictive immigration policies, which encompass ethnic, religious, and geographical exclusionary systems. These policies ensure that many refugees languish in Global South camps for generations, limiting their access to one of Canada's refugee resettlement programs.

2.3 *Canada's Refugee Resettlement Program*

Understanding the Canadian refugee resettlement system is essential for assessing refugees' vulnerabilities and requirements related to resettlement and mental health services. Table 1.2 shows the number of refugees arriving through the three Canadian refugee resettlement programs from January 2015 to October 2024, reflecting the current government policies on refugee quotas.

Table 1.2 Refugee numbers, Canada

Admission of resettled refugees by country of citizenship and immigration category January 2015–October 2024				
Country of origin	BVOR	GAR	PSR	Total
Syria	5650	52,475	43,560	101,685
Eritrea	595	4610	46,850	52,055
Iraq	435	6550	16,980	23,965
Afghanistan	85	30,730	29,545	60,360
Somalia	175	6605	11,395	18,175
Total	6940	100,970	148,330	256,240

Adapted from these sources: Government of Canada IRCC report (2024), Government of Canada, Statistics Canada (2022), Statistics Canada (2024a)

- Government-Assisted Refugee Program (GAR): Convention refugees referred by the UNHCR receive comprehensive resettlement assistance from the federal government.
- Private Sponsorship of Refugees Program (PSR): Convention refugees, sponsored by organizations and private individuals, receive full resettlement assistance from their sponsors.
- Blended Visa Office-Referred Program (BVOR): Convention refugees referred by the UNHCR receive 50% of their resettlement assistance from the federal government and 50% from their private sponsors.

Currently, 559 federally funded settlement organizations in Canada have agreements with Immigration, Refugees and Citizenship Canada to assist refugees (IRCC, 2024). In 2019, the Government of Canada led the way among countries by pledging to resettle more than 30,000 refugees annually. As shown in Table 1.3, that number declined significantly in 2020 with the start of the COVID-19 pandemic, doubled in 2021, and experienced a notable increase in 2022 and 2023 (IRCC-report, 2022).

2.4 *Resettlement Challenges*

Resettled refugees, in general, have high levels of stress as they transition from the precarious status of forced migration, where they have experienced extreme trauma, such as cultural oppression, gender-based violence, prolonged grief, and persistent psychological distress (Bokore, 2009, 2012, 2016; Kinzie, 2005; Levy & Sidel, 2006; Simich, 2008), to hope for a future in Canada. In addition, their accumulated stress compounds refugee health issues, including post-traumatic stress disorder (PTSD) and anxiety disorders (Bhui et al., 2006; Rees & Pease, 2007) (see Chaps. 7 and 8).

The experience of forced displacement and resettlement extends beyond emotional crisis to intergenerational trauma that impacts the well-being of survivors and their offspring. This ongoing crisis has a unique communal consequence affecting people economically, culturally, politically, and socially. It deregulates the normal function of a survivor's survival system for at least the next generation through trauma transference, creating disequilibrium between the body and brain, vulnerability to mental instability, and lack of emotional balance (Ager, 1999; van der Kolk, 2014).

Table 1.3 Number of resettled refugees in Canada

Number of resettled refugees in Canada by year					
2019	2020	2021	2022	2023	2024
30,080	9225	20,015	3770	51,000	51,081

Adapted from these sources: Government of Canada—Annual Report to Parliament on Immigration (2024); Government of Canada, Statistics Canada (2024)

What determines a successful integration of refugees is the accessibility and availability of support systems. The quality of the service they receive can do much to minimize or worsen emotional triggers from earlier stages of migration.

3 Summary

The challenge of finding durable solutions for refugees continues because protection and durable solutions for refugee crises are often limited to the host countries (e.g., Europe, rather than Africa). Even though the Global South hosts 80% of the world's refugees, issues of protection and the plight of refugees are rarely addressed in these countries' government policies or media. In contrast, in the Global North, political voices often oversimplify refugee issues, amplifying fear and the need for border control. Using terms such as *refugee crisis* and *migrant crisis* interchangeably confuses those who do not know the difference, impacting resettlement needs. Conflating the two is often designed to rally people behind the exclusion of those seen as the Other and hide the collective responsibility in the protection and integration of refugees. This political and systemic denial functions much like colonial/settler dynamics in Canada, in which the core humanitarian crisis experienced daily by many Indigenous groups is ignored, justifying inadequate support and shifting the blame from systemic problems to Indigenous groups.

Definitions

Durable solutions: Crisis intervention and long-term solutions based on voluntary repatriation, local integration, or resettlement as defined by the United Nations High Commission for Refugees. In the past 30 years, these solutions have become increasingly unfulfilled promises, leaving 75% of refugees in situations of protracted displacement for decades.

Migrant: A person who temporarily moves from one country to another in search of educational and economic opportunities to improve their life. They are known as lifestyle and labor migrants. The push-pull factor between the sending (home) and receiving (host) countries affects migrants.

Othering: According to Edward Said (1979), othering is based on race and geopolitical differences (us versus them) to prioritize the colonial culture over the colonized culture, which is cast as the inferior other. It has been perpetuated by Western explorers, poets, novelists, philosophers, political theorists, economists, and colonial administrators, beginning with Napoleon's occupation of Egypt in 1798. Concerning refugees, othering is present in politics, policy, and access to community services.

Refugee: A person forced to leave their home due to war and a threat to their life. They are protected under the 1951 UN Refugee Convention, or the Geneva Convention, signed by over 100 countries. Article 1A (2), Definition of the Term Refugee, defines a refugee as a person who, owing to well-founded fear of being persecuted for reasons of race, religion, nationality, membership of a particular social group, or political opinion, is outside the country of his nationality and is unable or, owing to such fear, is unwilling to avail himself of the protection of that country, or who, not having a nationality and being outside the country of his former habitual residence as a result of such events, is unable or, owing to such fear, is unwilling to return to it. (https://www.ohchr.org/en/instruments-mechanisms/instruments/convention-relating-status-refugees).

Discussion Questions

1. What are the human costs of the global refugee crisis?
2. Name the refugee-producing and hosting countries around the world. What are some causes of displacement in these refugee-producing countries?
3. How would you define Canadian refugee resettlement policy? What recent changes have been made to these policies?
4. What is the difference between the Government-Assisted Refugee (GAR), Private Sponsorship of Refugees (PSR), and Blended Visa Office-Referred (BVOR) programs? What are the service needs of each group?
5. Explain the impact of preferential treatment based on race in Western countries and its consequences for those forcefully displaced.

Competing Interests The author has no conflicts of interest to declare that are relevant to the content of this chapter

References

Ager, A. (Ed.). (1999). *Refugee perspectives on the experience of forced migration*. Pinter.

Bhui, K., Craig, T., Mohamud, S., Warfa, N., Stansfeld, S. A., Thornicroft, G., Curtis, S., & McCrone, P. (2006). Mental disorders among Somali refugees. *Social Psychiatry and Psychiatric Epidemiology, 41*(5), 400–408. https://doi.org/10.1007/s00127-006-0043-5.

Bokore, N. (2009). Female survivors of African wars dealing with the past and present. *Journal of Sociological Research, 1*(1). https://doi.org/10.5296/jsr.v1i1.189.

Bokore, N. (2012). Suffering in silence: a Canadian-Somali case study. *Journal of Social Work Practice, 27*(1), 95–113. https://doi.org/10.1080/02650533.2012.682979.

Bokore, N. (2016). *Somali-Canadian Women: Historical Past of Survival and Facing Everyday Challenges of Resettlement*. Available at: https://yorkspace.library.yorku.ca/items/ebfd63f6-9c7b-459e-9766-ff317e7ab348.

Dahlman, C. T. (2016). Unity amid barbed wire: asylum restrictions, European integration and the migration crisis. *Journal of Peacebuilding & Development, 11*(3), 8–22. https://doi.org/10.1080/15423166.2016.1222594.

Epp, M. (2017). *Refugees in Canada: A brief history*. Canadian Historical Association.

Government of Canada, Statistics Canada. (2022, October 26). *Ethnic or cultural origin by gender and age: Canada, provinces and territories, census metropolitan areas and census agglomerations with parts*. Retrieved March 26, 2023, from https://www12.statcan.gc.ca

Government of Canada: Canada. (2024). *A history of refuge*. Retrieved July 12, 2025, from https://www.canada.ca/en/immigration-refugees-citizenship/services/refugees/about-refugee-system/how-system-works/history.html

Government of Canada, Statistics Canada. (2024a, October 23). *Syrian refugees – Monthly IRCC updates*. Retrieved October 23, 2025, from https://open.canada.ca/data/en/dataset/01c85d28-2a81-4295-9c06-4af792a7c209

Government of Canada, Statistics Canada. (2024b, October 23). *A history of refugees – A timeline. Refugee milestones in Canada over the last 200 years*. Retrieved October 23, 2025, from https://www.canada.ca/en/immigration-refugees-citizenship/services/refugees/about-refugee-system/how-systemworks/timeline.html

Goździak, E. M., Main, I., & Suter, B. (Eds.). (2020). *Europe and the refugee response: A crisis of values?* Routledge.

IRCC. (2022). *Immigration, Refugees and Citizenship Canada, Permanent Residents – Monthly IRCC updates.* https://open.canada.ca/data/en/dataset/f7e5498e-0ad8-4417-85c9-9b8aff9b9eda.

IRCC. (2024). *A history of refugee in Canada.* https://www.canada.ca/en/immigration-refugees-citizenship/services/refugees/about-refugee-system/how-system-works/timeline.html

Isajiw, W. W. (1999). *Understanding diversity: Ethnicity and race in the Canadian context.* Thompson Educational Publishing.

Kagwanja, P. M. (2000). Ethnicity, gender and violence in Kenya. *Forced Migration Volume 9.* https://www.fmreview.org/kagwanja/.

Kinzie, J. D. (2005). Some of the Effects of Terrorism on Refugees. *Journal of Aggression, Maltreatment & Trauma, 9*(3–4), 411–420. https://doi.org/10.1300/J146v09n03_12.

Levy, B. S., & Sidel, V. W. (2006). *Social injustice and public health.* Oxford University Press.

Mathieu, S.-J. (2010). *North of the color line: Migration and Black resistance in Canada,* 1870–1955 (1st ed.). University of North Carolina Press.

Milner, J. (2009). *Refugees, the state and the politics of asylum in Africa.* Palgrave Macmillan.

Milner, J., & Wojnarowicz, K. (2017). Power in the Global Refugee Regime: *Understanding expressions and experiences of power in global and local contexts. Refuge, 33*(1), 7–17. https://doi.org/10.25071/1920-7336.40444.

Papadopoulos, R. (Ed.). (2002). *Therapeutic care for refugees: No place like home* (1st ed.). Routledge, Karnac Books.

Papadopoulos, R. K. (2006). Refugees and psychological trauma: Psychological perspective. Research Gate. http://isites.harvard.edu/fs/docs/icb.topic920418.files/arc_1_10refandpsych-1.pdf

Papadopoulos, R. K. (2021). *Involuntary dislocation: Home, trauma, resilience, and Adversity-Activated Development.* Routledge, Taylor and Francis.

Paynter, E. (2022). Border crises and migrant deservingness: How the refugee/economic migrant binary racializes asylum and affects migrants' navigation of reception. *Journal of Immigrant & Refugee Studies, 20*(2), 293–306. https://doi.org/10.1080/15562948.2021.1980172

Rees, S., & Pease, B. (2007). Domestic Violence in Refugee Families in Australia. *Journal of Immigrant & Refugee Studies, 5*(2), 1–19. https://doi.org/10.1300/J500v05n02_01.

Said, E. W. (1979). *Orientalism.* Vintage.

Said, E. W. (1993). *Culture and imperialism.* Vintage.

Simich, L. (2008). *Refugees, mental health and the meaning of "Home".* http://www.metropolis.net/pdfs/fow_CITC_fall2010_e.pdf

Statistics Canada. (2025). *Immigration and ethnocultural diversity statistics.* Retrieved July 12, 2025, from https://www.statcan.gc.ca/en/subjects-start/immigration_and_ethnocultural_diversity

The Canadian Encyclopedia. (2025). *Timeline of immigration.* Retrieved July 12, 2025, from https://www.thecanadianencyclopedia.ca/en/timeline/immigration

UNHCR global trends - forced displacement in 2021. (2025, October 23). UNHCR Flagship. https://www.unhcr.org/media/global-trends-report-2021.

UNHCR global trends - forced displacement in 2022. (2025, October 23). UNHCR Flagship. https://www.unhcr.org/global-trends-report-2022.

UNHCR global trends - Global Trends report 2023, (2025 October 23). UNHCR Flagship. https://www.unhcr.org/media/global-trends-report-2023.
UNHCR global trends - Global Trends report 2024, (2025 October 23). UNHCR Flagship. https://www.unhcr.org/global-trends-report-2024.
Van Der Kolk, B. A. (2014). *The body keeps the score: Brain, mind, and body in the healing of trauma*. Penguin Publishing Group.
Whitaker, R. (1991). *Sovereign Idea: Essays on Canada as a Democratic Community*. McGill-Queen's Press.
Whitfield, H. A. (2006). Blacks on the Border: *Black Refugees in British North America*, 1815–1860. Burlington: University of Vermont Press.

Chapter 2
Trauma and Stages of Migration

1 The Three Stages of Forced Migration

The multidimensional cause of human displacement is discussed in Chap. 1. For those in the Global South, forced migration is caused by an internal conflict (civil war, such as the Syrian conflict for regime change that ended in late 2024, and internal struggles, such as the fallout of the Arab Spring in late 2010, or external war (invasion or occupation by an external power, such as the US invasion of Iraq). In the global refugee crisis of the past few decades, where human displacement is becoming all too familiar, understanding what happens to the survivors of these conflicts is critical for social work practice. The increased number of refugees in the world—more than 100 million—demands that social work pay attention to the need for adequate services. These services begin with social workers having a basic understanding of the impact of conflict and war on those who are forcibly displaced from their homes because of a threat to their lives and their resettlement in Canada.

The 1951 UN Refugee Convention defines refugees as those who, under threat, leave their homes, escaping death or political oppression. Unfortunately, refugees, as defined by the Convention, are sometimes confused with economic migrants who arrive at international borders seeking a better life. The confusion means refugees resettling are denied access to necessary services. The lack of services further impacts resettled refugees who are affected by their past trauma, influencing their safety and resettlement outcomes (Bokore, 2016).

A common framework used to understand refugees' experiences and trauma is the three stages of forced migration (Ager, 1999; Papadopoulos, 2021; Potocky-Tripodi, 2002): pre-migration, transit, and resettlement. Pre-migration experiences begin before war erupts. At this stage, survivors face uncertainty, fear, and emotional distress that come with the anticipation of danger and lack of safety (Ager, 1999; van Der Kolk, 2014). Their anxiety activates the autonomic reaction of the sympathetic nervous system (fight, flight, or freeze; see Chap. 3). This response

N. Bokore, *Trauma-Informed, Culturally Based Intervention*,
https://doi.org/10.1007/978-3-032-06835-4_2

continues during the transit stage when the individual (or family) moves to an interim country to seek protection and safety under the aegis of the UNHCR. The uncertainty continues here as the host country may or may not accept refugee claims, and refugees, particularly women and children, lack basic needs such as food, shelter, and safety (Ager, 1999; Bokore, 2009). Each migration stage creates a stored memory, determining refugees' responses to future stressors and impacting their well-being (Bokore et al., 2023; Levine, 2015; van Der Kolk, 2014). For survivors, these initial stages create a frame of reference for their future reactions (see Chaps. 3 and 4). Depending on their framework (past experiences), some refugees build resilience, overcoming fear, uncertainty, and hopelessness (see Chap. 4). Others are trapped in cycles of fight, flight, or freeze (sympathetic, or dorsal vagal, activation), negatively impacting their long-term health and well-being (van Der Kolk, 2014; see Chap. 3).

Refugees' resettlement experiences in a host country (stage three), often a Western immigrant-receiving nation, vary. However, despite these variations, forced migration is devastating and emotionally overwhelming for survivors as they deal with the loss of their homes, community safety nets, families, identities, and country (Ager, 1999; Papadopoulos, 2002, 2006). The emotional response to this devastation continues during resettlement, when refugees are confronted with new stressful challenges such as accessing resources and making social connections (see Chap. 3).

The personal stories of a few refugees who participated in my research will be discussed below. These stories show their constructions of events and meaning-making process as they navigate their displacement and later resettlement issues. Their stories recall moments of fear and overwhelming stress. Some felt helpless and disoriented about what was happening; others were frozen and unable to move. Still others describe their emotional struggles, feeling overwhelmed and emotionally shut down. These accounts do not give a complete picture of the cause and extent of trauma but reflect the thought processes and experiences of refugees at each migration stage.[1]

2 Pre-Migration (Stage One) Stories of Survivors

In the pre-migration stage, refugees often experience job insecurity and anticipate a disruption to their income generation because of the impending conflict or war. This economic uncertainty affects families and communities, and in tandem with rising political tension, can increase their stress level, creating mistrust and loss of social connections essential to who they were as members of a communal society in times

[1] These stories and those later in the chapter are drawn from focus groups I conducted for a research project on trauma-informed, culturally based intervention (TICBI) during the COVID-19 outbreak. There were a total of 22 participants who had resettled in Canada from war-affected countries: Somalia, Ethiopia, Iraq, Syria, Yemen, and Sri Lanka. See Bokore et al. (2023).

of peace. Refugees sometimes look back at this stage through nostalgic pre-war memories while describing the fear they experienced in each stage. Communal community support systems, which functioned as a buffer when hardships set in and safety was needed, are eroded in this stage. At this stage, many refugees are forced to make compromises to access resources and information as war looms. Heads of families feel extreme pressure to ensure the well-being of their family, which continues during the second, or transit, stage of forced migration.

Personal experience: My family was forcibly displaced after the military government in Ethiopia began to identify members of prominent families as being a threat to the country's security, particularly those with Somali, Eritrean, or Oromo ethnic backgrounds. The government empowered local community police, known as *Kebele* (ward) members, to imprison anyone who opposed the government. My two colleagues and I were imprisoned and tortured, which still impacts my health. I was fortunate to be released after seven months, but my two colleagues died in prison. After my release, escape became my only option. That is when I decided to cross the first border with my children into Djibouti, seeking asylum. Due to the inhospitable weather conditions for homeless refugees, I crossed the second border into Somalia. I crossed my final Mediterranean border to Italy before arriving in Canada through one of the Canadian humanitarian programs. The following quotes were taken from my research participants discussing their experiences at this stage.

A Syrian resettled refugee said: After the Arab Spring, there were fears of government change and sectarian violence, which is what my family and I were concerned about, and when the violence began, we had to leave…. Due to safety concerns on the road, we had difficulty deciding which country to go to if we needed to escape and how to get there.

An Iraqi resettled refugee said: This stage was stressful for me due to the impending war of the American invasion. Following the American and other Western media, I knew they were going to invade our country. I experienced fear of the unknown. As a result, my family and I did not know where to go or what to do. I guess the fear was coming from our past experiences with the Iran and Iraq wars [1980–1988]. We were familiar with regional wars and knew about safe places, but the American invasion differed from the others. We feared its impact and how we all were at risk, with nowhere to hide. Another thing is that my dad had businesses in different cities, and we were concerned about losing all of them at once and what would happen to our family after that.

A Somali resettled refugee said: One day, a family member asked me to get her sugar from the store. Some military men who I think were related to Siyad Barre were standing in front of our door, and as soon as I came out, they asked me to identify myself by my tribe. I could not tell them about my tribe because of my accent—I was afraid they would not believe me because, at the time, they were killing tribes from the north who spoke as I did. I almost died that day. Fortunately, my family heard my screams and rescued me, explaining my tribal connection to eastern Ethiopian-Somalis, not Somaliland tribes.

A Sri Lankan resettled refugee said: I was born and lived in a war zone until I was nine. So, every day was traumatic for me because there was no security in my area. When the army or the military moves to your city or neighborhood, since you are Tamil, they will look at you as their enemy. Their only intention is to hurt you or kill you. Their job is to go after the men in the family, and they will execute them. Whether they were part of the rebels or the revolutionary movement did not matter. As a Tamil boy between 12 and 13, I was a target for them. It was scary. They will select boys to hit, put them in jail, or do whatever they decide to do. So, I grew up in fear until we left.

3 Forced Migration (Stage Two or Transit-Stage) Stories

Personal experience: The fears, anxiety, and mistrust of my family that began during pre-migration continued into the transit, or devastating, stage. I remember our flight out of Ethiopia to Djibouti. We could not stay in Djibouti because of systemic challenges in the refugee registration system and extreme heat, so we crossed the border into Somalia. As a young, unprotected woman, Somalia posed its own dangers for me, including gender-based violence and difficulty feeding, educating, and protecting my children. With the start of the internal political struggle in Somalia that resulted in the collapse of the Somali government in 1991, we were again at risk, from which the UNHCR could not protect us. Because I saw in Somalia a parallel beginning of the internal political conflict in Ethiopia, and trusting my journalistic instinct and training, I decided to escape from Somalia, crossing the Mediterranean to Europe. As a mother with four young children who, in their short lives, had witnessed extreme horror and senseless violence, my vulnerability was acute until we arrived in Italy. We received a warm reception, typical for Ethiopian refugees at the time, because of our shared history with Italians during the First and Second World Wars and their invasion of Ethiopia. We received shelter from religious organizations and food from Caritas Italiana, and the World Council of Churches (WCC), working with the UNHCR, provided us with ID and registration to resettle in a third country. Despite these support systems, the impact of the first two stages of migration affected me and my family emotionally. As a family, we sometimes struggle because our overactive, hypervigilant minds keep us alert for danger, causing unnecessary stress. The following quotes are taken from research participants,[2] who shared their stories about their experiences at this stage with me.

A Somali-Ethiopian story: I was displaced twice. The first time was during the 1977–1978 Somali-Ethiopian war when I escaped to Somaliland. I went back to Ethiopia when the war ended [only] to move back [to Somaliland] again when the military government of the time started targeting people in my community. I still feel like I have been running for most of my adult life, becoming either a refugee in another country or an IDP. My worst memory of this journey is from 1988, when I gave birth to my first child while running from Ethiopia to Somalia. After I resettled

[2] The research was cleared by Carleton University's Research Ethics Boards (CUREB)—A.

in Canada, I had three more children, and remembering the past, I was preparing each time for my worst childbirth experiences to happen again.

The story of a Somali from the Somali Republic: Arriving at camp was confusing, and being there the first year was difficult. Everything was hard. We were afraid for our missing families and friends, and we did not know at the time where they were. I was close to death multiple times and separated from my family numerous times, and came close to being attacked once, but what saved me until I arrived in Canada was my spiritual and cultural strength, resiliency, and survival instinct. However, I need to find a solution to my anxious feelings and fear of danger that sometimes come up.

A Syrian story: I was separated from my family and felt isolated and lost until I reunited with some of them at a camp on the border with Turkey. I lost family members before I left my country and during the journey to the camp. I was mourning those I lost before and after arriving, and I still do. Consistent religious prayers and reading the Qur'an saved me after losing so many friends, family members, and neighbors in Syria due to the bombings. I still think about how close I came to losing my life and how I escaped those dangerous events. It is hard to trust people after what I saw, but I often pray, which helps. Religion kept me alive and gave me strength while dealing with the fear of receiving the ever-present unpleasant news in a refugee camp.

An Iraqi story: I experienced war violence that was physical and mental during my journey and at the camp in Jordan. I was affected by hearing stories about how my brother died after I left. I was in bad shape, but my family's support and faith saved my life. It also helped me cope with watching my parents' heartbreak after the loss of my brother, who was their older son, and the stress of pretending to be strong for them. My religious beliefs enabled me to cope with it all. We all saw things that we cannot even describe. Nevertheless, what we have is our faith.

4 Resettlement (Stage Three) Stories

Even though resettled refugees still think about the first two stages with a sense of fear, the third, or resettlement stage, is where the support quality determines the success of refugees' integration into their host country. My success story is being able to raise my children, who are living comfortably in Canada, and my academic journey that led me to write this book. The following are the stories of my participants.[3]

A Syrian story: Finally, being resettled in Canada was a positive outcome for my family. As a man, I am still struggling but staying strong for myself and my family. I am a privately sponsored refugee. However, I have friends who are

[3] The study this book is based on was cleared by the Carleton University's Research Ethics Boards (CUREB)—A.

government-sponsored refugees, and they are telling me it is worse for them because 12 government-assisted refugee families may have one counselor assisting them with resettlement needs. [In contrast,] for those privately sponsored like me, twelve sponsors are helping one refugee family, which is helpful for resettlement. I think privately sponsored refugees find jobs more efficiently than GARs. They tend to get more support—financially and emotionally. However, I sometimes feel sponsorship is like ownership, which should not be, but I feel obligated to do what my sponsor tells me to do. Before and after the uprising, many Syrians relied on religion and herbal treatment for various illnesses, including stress or sadness. This practice and our tight-knit, family-oriented group help me with my emotions when I struggle.

An Iraqi story: Finally, getting resettled in Canada was suitable for my family, but as a man, I am still trying to stay strong for myself and my family. I needed services that would help me understand my situation and emotions and support me with clear communication and information. I was confused most of the time by the personal questions [of service providers], but I needed to understand their culture. I was a religious person before I came, and I still use my spirituality, prayers, patience, and talking to elders to help me with my emotions. Our culture tells us to work together as a community to save our families and survive challenging times. That is what I am mostly doing at the moment.

A Somali story: I came alone with private sponsorship. I miss the family I left behind in the camp, including my parents. I kept calling them because they did not have the Internet; I always told them I missed being with them there. I was fortunate to be a neighbor to someone who works in the Somali center, and they helped me with many things, and my brother also helped me. My friend and I found housing in the same neighborhood when we came to Canada. We came from two different tribes, but we started talking when we met and immediately became friends. We helped each other and overcame the tribal division created back home and in the diaspora. We focused on our everyday needs as mothers and newcomers, adapting to a new country for us and our children. Our friendship became a lasting, supportive system for each other and our children.

An Ethiopian story: In my country, we have a circle of support—but I do not have it here. I feel lost. I felt isolated because I lacked an understanding of the new, divisive culture in my community. The combined loss of community and culture is creating emotional challenges for me. Some service providers never ask about what I am going through emotionally, my cultural values, or my concerns about mental health-related conversations. Prayer is what helps me mainly to cope with the past and present challenges. Yes, prayers and culturally based socialization with some friends over coffee still help. During those challenging years, I also did not lose hope, and I sometimes thought that something bright was out there for me and that I would find it and be saved.

A Sri Lankan story: I was a child when we came here, but I always felt my mom's fear. I have often felt fearful at times over the past eighteen years. I do not know why I am still afraid of fights or arguments. After my arrival, as a refugee child living in a predominantly Caucasian and blue-collar neighborhood, I struggled with isolation. I felt isolated because I lacked support and dealt with my emotional challenges alone. Some service providers never ask about past or present trauma. My family and a solid connection to nature and religion were my sources of survival. I think that this is what keeps me going through life.

5 Emotional Responses: Three Stages

As these refugee accounts attest, refugee families, unlike immigrants, arrive in their countries of resettlement having experienced traumatic, life-changing events throughout their migration. People respond to these events in unique ways. Some continue their lives without emotional disruption, having the ability to reappraise their experiences positively, adopt coping strategies, and work at post-traumatic growth (see Chap. 4). However, many experience the adverse effects of trauma—anxiety, terror, shock, shame, depression, intrusive thoughts, helplessness, powerlessness, emotional numbness, and disconnection—with lasting consequences (Ehntholt & Yule, 2006; Evans & Coccoma, 2014; Han, 2015; see Chap. 3). For this latter group, the added stress of resettlement weakens their sense of control, connection, and meaning, which, in turn, affects their well-being and continues into the next generation through trauma transference (Cozolino, 2010; Siegel, 2011). This book is about understanding the sources of trauma, planning appropriate interventions, and providing them effectively (see Chaps. 7–9).

Stage one of forced migration is when anxiety, fear, and anticipation about the unknown set in for refugees (Ager, 1999; Papadopoulos, 2006, 2021). This creates overwhelming stress and activates body and brain responses that stimulate the sympathetic nervous system's survival reaction of fight, flight, or freeze.

Stage two entails traumatic events that happen when refugees are en route to an interim, or refugee-hosting, country for safety. Depending on the severity of their journey, refugees' fear-filled responses continue during this stage. Furthermore, the trauma continues in this period—more violence and loss—which survivors carry with them into resettlement. As a result of this violence, emotions become stuck points, often creating maladaptive responses such as avoidance or extreme hypervigilance, causing isolation and affecting refugees' capacity for social interaction and establishing the connections they need for resettlement (Ager, 1999; Levine, 2010, 2015; Papadopoulos, 2006, 2021; Porges & Dana, 2018; van Der Kolk, 2014).

For refugees, physiological change and emotional dysregulation are part of their migration as they are immersed in stressful environments. Nevertheless, in stage three, survivors feel safe enough to establish social connections and integrate into a resettlement country in the West. However, the traumatic events that occurred during their forced migration shape their emotional being, affecting their ability to feel safe and establish new relationships in resettlement countries, which in large part determines the success of integration (Chaps. 3 and 4)—resettlement challenges may still trigger them. Then their pre-migration history reinforces these challenges, including how they participate professionally, socially, and politically in their new home. Those actively involved in political struggles in their home country, which were the source of their displacement, may face emotional challenges as they decide whether to continue their political activism during resettlement. Socially, some fear for the safety of family members back home because of the presence of informants within their refugee communities reporting back to their home country's government. Some immerse themselves in self-defeating thoughts arising from past struggles and entrench adaptive survival behaviors. This leads to misinformation and misconceptions about the resettlement country (e.g., fear that religious and cultural change will be imposed on them). Hence, they avoid social interaction, turning to self-imposed isolation instead.

When trauma happens, human emotions become complex and difficult to describe objectively because of the multiple factors that determine the response (Bokore et al., 2023). For example, when I sense negative energy, memories from my past draw me to the possibility of pending danger. It comes from implicit, or involuntary, physical sensations initiating these sensations in me. The emotional legacy of war stayed with me decades after my initial traumatic memories were formed. Some of those memories still haunt me, becoming my mental models, or what Daniel Siegel (2011, pp. 151–152) calls "schemas," which I use to assess the level of danger I am in or to self-regulate my emotions until the feeling of threat passes. Depending on refugees' resettlement experience, their emotional hypersensitivity, or over-reactivity, to stressful situations creates toxic living conditions that affect their well-being and that of their families. They live with a constant emotional activation that causes physical exhaustion and health issues.

In summary, the traumatic events during the three stages of forced migration, including losses refugees experience during stage two, overwhelm some survivors, affecting their adaptation during the third stage, where having a sense of emotional control, connection, and meaning-making is vital to successful resettlement (van Der Kolk, 2014).

Definitions

An ethnic conflict as a source of displacement: This is a manufactured crisis often caused by the leaders of ethnic groups competing for power, access to resources, and territory. Ethnic leaders influenced by foreign powers with political agendas play a crucial role in mobilizing the groups engaged in ethnic conflicts.

Forced migration refers to the involuntary relocation of people from their homeland or the place they call home. This phenomenon, also known as displacement, can be stressful, leading to psychological crises like PTSD for some individuals, while providing opportunities for change and growth for others.

Refugee trauma: This is often associated with the feeling of being in a war zone where your life is threatened. It is an experience that significantly increases the risk of developing dysfunctional and maladaptive behaviors, as well as other trauma-related health disorders such as PTSD. The development of these issues heightens individual crises that alter people's perceptions of others and their environments, adding to their psychological and emotional wounds.

Discussion Questions

1. What are refugees' emotional responses to the stages of forced migration?
2. When reading the stories of survivors during transit or the second stage of migration, what stands out to you?
3. What are the differences between the survivors' experiences at each stage of migration and their everyday experiences that create emotional challenges during each stage of forced migration?

Competing Interests The author has no conflicts of interest to declare that are relevant to the content of this chapter.

References

Ager, A. (1999). *Refugees: Perspectives on the experience of forced migration*. Pinter.

Bokore, N. (2009). Female survivors of African wars dealing with the past and present. *Journal of Sociological Research, 1*(1). https://doi.org/10.5296/jsr.v1i1.189.

Bokore, N. (2016). *Somali-Canadian Women: Historical Past of Survival and Facing Everyday Challenges of Resettlement*. Available at: https://yorkspace.library.yorku.ca/items/ebfd63f6-9c7b-459e-9766-ff317e7ab348.

Bokore, N., McGrath, S. L., McGuire, P., Karod, A., Rahimpour, M., & Asokumar, A. (2023). Developing a trauma-informed culturally-based intervention (TICBI) approach for refugee resettlement practices. *International Journal of Social Work, 10*(1), 1. https://doi.org/10.5296/ijsw.v10i1.20561

Cozolino, L. J. (2010). *The neuroscience of psychotherapy: Healing the social brain* (2nd ed.). W. W. Norton & Company.

Ehntholt, K. A., & Yule, W. (2006). Practitioner review: Assessment and treatment of refugee children and adolescents who have experienced war-related trauma. *Journal of Child Psychology and Psychiatry, 47*, 1197–1210. https://doi.org/10.1111/j.1469-7610.2006.01638.x

Evans, A., & Coccoma, P. (2014). *Trauma-informed care: How neuroscience influences practice*. Routledge.

Han, S. (2015). Understanding cultural differences in human behavior: A cultural neuroscience approach. *Current Opinion in Behavioural Sciences, 3*, 68–72. https://doi.org/10.1016/j.cobeha.2015.01.013

Levine, P. (2010). *In an unspoken voice: How the body releases trauma and restores goodness*. North Atlantic Books.

Levine, P. (2015). Trauma and memory: Brain and body in a search for the living past. *A practical guide for understanding and working with traumatic memory*. North Atlantic Books.

Papadopoulos, R. K. (2002). *Therapeutic care for refugees: No place like home* (1st ed.). Routledge.

Papadopoulos, R. K. (2006). Refugees and psychological trauma: Psychological perspective. *Research Gate*. http://isites.harvard.edu/fs/docs/icb.topic920418.files/arc_1_10refandpsych-1.pdf.

Papadopoulos, R. K. (2021). *Involuntary dislocation: Home, trauma, resilience, and Adversity-Activated Development*. Routledge, Taylor and Francis.

Potocky-Tripodi, M. (2002). *Best practices for social work with refugees & immigrants*. Columbia University Press.

Porges, S. W., & Dana, D. (2018). *Clinical applications of the polyvagal theory: The emergence of Polyvagal-Informed therapies*. (Norton Series on Interpersonal Neurobiology). W. W. Norton & Company.

Siegel, D. J. (2011). *Mindsight: The new science of personal transformation*. Bantam.

van Der Kolk, B. A. (2014). *The body keeps the score: Brain, mind, and body in the healing of trauma*. Penguin Publishing Group.

Part II
The Neurobiology of Trauma

Chapter 3
The Development of Post-Traumatic Stress Disorder (PTSD)

1 Introduction

The human brain is prepared for survival and sustaining life. During a traumatic event, such as a threat to survival, the brain and the body respond until the threat passes. Unfortunately, for some survivors, without help, threat responses continue, creating a prolonged cycle of traumatic memories that keeps them trapped, adversely affecting their well-being.

For those providing trauma-informed, culturally based intervention (TICBI), understanding survivors' responses is the first step in helping (Bokore et al., 2023). Trauma disrupts the connection between the hypothalamus and the rest of the limbic system that controls the autonomic nervous system, which includes the sympathetic nervous system (SNS) and parasympathetic nervous system (PNS). Trauma affects individuals, families, and communities as it decreases the functions of people's PNS and overactivates their SNS (LeDoux, 1996; van der Kolk, 2014).

In some communities and ethnic groups, the impact of the prolonged dysfunction of the PNS and SNS becomes multi-generational, creating maladaptive responses to life stressors even after resettlement as a form of "insidious trauma" (McLean et al., 2015, p. 365)—the cumulative effect of interpersonal and institutional social and cultural/religious oppression and discrimination. Brave Heart et al. (2015) take a close look at the historical impact of insidious trauma on Indigenous groups in North America: the loss of land and resources, the fragmentation of families and tribal groups, the impact of residential schools, the banning of ancient spiritual/healing practices, and the destructive behaviors that result, such as the use of alcohol and other substances. Van der Kolk (2014) examines the effect of cumulative trauma on the well-being of individual survivors of war and their communities. He defines trauma as a life-changing effect that has devastating health consequences caused by behavioral, psychological, and cognitive disturbances. Similarly, other researchers have explored the neurobiological impact of trauma that lasts for

N. Bokore, *Trauma-Informed, Culturally Based Intervention*,
https://doi.org/10.1007/978-3-032-06835-4_3

generations due to the ongoing dysregulation of the SNS and PNS, putting individuals and communities at risk (Bokore, 2019; Michopoulos et al., 2016; Ona, 2020). For survivors, as their brain defense mechanisms (SNS) remain activated by day-to-day events or perceived life-threatening situations, their body/brain stays alert even when the threat has been removed and the person is in a safe environment (decreased PNS activation), creating constant emotional dysregulation (Brave Heart et al., 2015).

While there are positive psychological, behavioral, and cognitive responses to trauma and adaptive states for some (Bokore et al., 2023; Tedeschi & Calhoun, 1995, 1996; see Chap. 4), many survivors suffer from emotional distress, hypervigilance, intrusive memories, and flashbacks (Hinton & Good, 2016). For some refugees, their traumatic experiences change their brain and body responses to stress, creating deep and enduring wounds that can sometimes last for generations. Some of these individuals engage in maladaptive behaviors or activities involving drug and alcohol use or increased food consumption, which further impact their well-being (McCauley et al., 2012; Siegel, 2011) as they continue to participate in these risky behaviors affecting their lives and their children's (Blume et al., 2000; McCauley et al., 2012; Sareen, 2014; Volpicelli et al., 1999).

2 Causes of PTSD

Before diving into the risks of developing PTSD, let us look at the cause. As mentioned above, war and conflict cause reactionary psychological and biological changes in people, often exhibiting as PTSD, a mental health condition triggered by life-threatening or terrifying events. Even though the impact of trauma has been known in veteran communities since the end of the Second World War, PTSD was only added to the *American Psychiatric Association's Diagnostic and Statistical Manual* (DSM-III) in the 1980s following advocacy by service providers working with veterans of the Vietnam War who saw the struggles they endured without receiving adequate services (Evans & Coccoma, 2014, p. 5).

With the emergence of neuroscience research in the 1990s, the impact of trauma, including PTSD, was further explored and people's hyperresponsivity to threats created by alterations in the function of the brain and body (van der Kolk, 2014). Since then, the many faces of trauma and the development and manifestation of PTSD have been studied by various researchers, detailing the long-term effects of biological, physical, and emotional changes (Cozolino, 2010; LeDoux, 1995, 1996; Levine, 2015; Ogden et al., 2006; Porges, 2007; van der Kolk, 2014; Wilkinson, 2010). Survivors of wars and conflicts also experience uncontrollable levels of anxiety and fear, especially prevalent in refugee communities (Simich, n.d.). These symptoms can be exacerbated by a lack of support systems during forced migration and increased exposure to highly toxic environments.

Both the *International Classification of Diseases* (ICD-11) *and the Diagnostic and Statistical Manual of Mental Disorders* (DSM-5) list the symptoms of PTSD. The difference is that the ICD-11 classifies PTSD as a mental, behavioral, and neurodevelopmental disorder (Ben-Ezra et al., 2020; Schaffer & Rodolfa, 2017) and the DSM-5 defines PTSD as a trauma- and stress-related disorder that manifests itself in intrusive thoughts, avoidant behaviors, cognitive effects, and alterations in the arousal and activity of the autonomic systems (Brewin et al., 2010, pp. 367–368). The World Health Organization (WHO, 2020) describes the emotional impact of PTSD as feeling worthless, being afraid, a diminished belief in personal ability to cope with life stress, shame and guilt, self-isolation, the inability to regulate emotions, and difficulty moderating emotional responses.

Some researchers argue that the symptoms listed in both the ICD-11 and the DSM-5 are Western-based and call into question their suitability for cross-cultural interpretation and use in intervention (de Jong et al., 2003; Jones et al., 2003; Kienzler, 2008; Kleinman & Kleinman, 1991). Despite this debate, the consequences of PTSD are noted by survivors as their neurophysiological reactions change, triggered by their frequent fight, flight, or freeze responses. Likewise, service providers have noted PTSD symptoms as they deal with interpersonal, psychosocial, and physical issues, and the emotional cost of family breakdowns, evidenced in aggression and intimate partner violence (IPV), divorce, neglect of children, and difficulty raising emotionally healthy children.

3 The Prevalence of PTSD

In exploring PTSD and its prevalence, it is essential to talk about the causes and experiences of trauma. The word "trauma" originally comes from the Greek word "*trauma*," meaning wound. Trauma can thus be understood as an invisible wound causing multiple health risks, such as PTSD (LeDoux, 1996; van der Kolk, 2014). For survivors of wars and conflicts, persistent exposure to violence creates an emotional wound. Various disciplines, such as social work, psychology, and neuroscience, provide definitions of trauma and address its impact (Knight, 2015; Ogden & Fisher, 2015; van der Kolk, 2014). Trauma is a known human response to distressing physical and/or emotional events and can impact individuals regardless of gender, age, race, social location, or sexual orientation (Kessler et al., 2017; van der Kolk, 2014).

In the era of social media, the definition of trauma has become even more varied, where the words *stressful* and *traumatic* are used interchangeably to address minor day-to-day challenges. However, in the context of widespread war, trauma has a more profound meaning as a life-changing event with painful memories (Bokore, 2017; Bokore et al., 2023). For some survivors of such events, memories affect their capacity to cope, defining their narrative of survival, self-concept, relationships, and evaluation of future life-threatening conditions. For survivors who develop PTSD, the overwhelming implicit memory of their experiences continues to define their reactions to current events, depriving them of the ability to establish meaningful relationships or enjoy life.

With advancements in social neuroscience, researchers like Joseph LeDoux have begun investigating the neurobiological and psychological processes involved in the development, maintenance, and recovery from PTSD and other mental illnesses (LeDoux, 1995, 1996). Daniel Siegel (2007, 2011) explores the responses of brain regions involved in conditioning, habituation, stimulus generalization, resistance, and other impaired functions related to healing. Bessel van der Kolk (2014) focused on the body–brain connection in trauma response and healing practices. Others, including Louis John Cozolino (2010), outlined the functions of PTSD and the effects of dysregulation in the underlying neurocircuitry, which influences cognitive-emotional interactions and threat appraisals, increasing survivors' vulnerability to PTSD as well as intervention practices.

4 The Neurobiological Process of PTSD

Trauma overwhelmingly affects the brain, particularly the amygdala (the fear center), hippocampus (which is involved in learning, memory, and the development and experience of trauma-related mental disorders, especially post-traumatic stress disorder), and prefrontal cortex (responsible for decision-making, rational thought, and emotional regulation). As a result, these regions respectively regulate stress, memory, and emotional control. Trauma can lead to dysregulation and lasting changes in their function and structure. This may manifest as increased reactivity to perceived threats, difficulty controlling emotions and impulses, and impaired memory. The neurobiological effect of trauma on the brain starts with the activation of the fear circuit, causing brain stress and dysfunction. This dysfunction starts a chain reaction in the brain, changing both the function (chemical processes) and structure of the brain, particularly in the areas that deal with stress (hypothalamic-pituitary-adrenal), fear (amygdala), and learning and memory (hippocampus). The amygdala uses a relay system (thalamus) to share incoming threat and related information to other areas, connecting the brainstem, cerebral cortex, midbrain, and medial prefrontal cortex (Lanius & Olff, 2017; LeDoux, 1996).

Experiencing conflict/war and displacement creates a hyperactive stress response involving the constant activation of the hypothalamic–pituitary–adrenal (HPA) gland. The activation of this area plays a critical role in releasing the stress hormone corticotropin to stimulate the release of the adrenocorticotropic hormone (ACTH) from the pituitary gland, allowing the survivors to manage stress and plan their escape (LeDoux, 1996). The HPA then activates the autonomic nervous system (ANS) for a rapid response, promoting the physiological changes needed for synaptic transmissions of the sympathetic and parasympathetic nervous systems and their distinct roles for (fight or flight and feeling safe) systems during escape. The two areas, the HPA axis and the ANS, function as the key organizers of the survivor's response to traumatic stress and the development of PTSD (Fillenz, 1990; Tache & Brunnhuber, 2008). Even though symptoms of PTSD are precipitated by stress (neuroendocrine response), not all survivors are affected by the reactions following extreme stress (see Chap. 4). However, for those who are, trauma becomes life-changing and intergenerational.

Traumatic events activate the amygdala (fear circuit; Levine, 2015; Porges, 2007; Seigel, 2011), also known as the "smoke detector" (van der Kolk, 2014, p. 98). It is located in the medial temporal lobes on both sides of the brain and actively responds to threats or biologically relevant environmental stressors. The parietal lobe's sensory cortex works with the amygdala to make sense of incoming information (vision, sound, smell, taste, and touch) from various body parts. It is a lifesaving system (life-changing when it comes to trauma) that develops during the final months of gestation and the first few years of a child's life. Studies on this part of animal brains have provided knowledge about microanatomy and the cellular function of this area during fear or extreme stress. Recent human studies using functional imaging have contributed further knowledge about its structure and function in human-conditioned fear in conscious and unconscious responses. The amygdala triggers defensive behaviors of fight, flight, or freeze (immobility), triggering the endocrine system's autonomic activation. It also initiates physiological changes, increasing the blood flow and energy supply to skeletal muscles, providing survivors the chance for escape and survival. The locus coeruleus noradrenergic system (LC-NE) throughout the central nervous system also has a role in these lifesaving responses, modulating the amygdala's fear and stress-induced behaviors. When damage occurs to these coordinated areas, the visible signs are emotional dysregulation, difficulty in problem-solving, poor judgment or decision-making, and distorted emotional memories affecting social life (Dana, 2020).

The human brain coordinates brain response and survival during trauma. This makes it necessary when working with survivors who have PTSD to understand, for example, how the locus coeruleus (LC), a small structure connecting the brainstem, midbrain, and cerebral cortex, modulates a diverse range of stress and trauma-related behaviors, including the arousal state. The LC generates norepinephrine (NE), a neurotransmitter and a chemical excreted during stress, signaling a response across the nerve endings to cells, including muscle glands, increasing the chance of escape and survival (Cozolino, 2010). Norepinephrine is one of the main components of the emergency function in the sympathetic nervous system's response (fight or flight). There is little evidence directly linking damage to the LC and trauma responses such as PTSD. However, the hypothesis is that a deregulated LC affects cellular functions, causing sudden memory loss, anxiety, an increased startle response, hallucinations, delusions, and depression, all of which are trauma responses (Hansen, 2021; Naegeli et al., 2018).

One of the brain's efficient mechanisms during trauma responses, such as PTSD, is its capacity for connectivity. The efficiency comes from the brain's central relay system, which is managed by the thalamus, also known as the "customer service center." Located deep within the brain's center, it analyzes the raw data collected from incoming body/brain responses and threat sensations (Cozolino, 2010). Its sensory-specific function allows for the quick analysis of the raw data, relaying it to other main areas in trauma response, such as the amygdala and hippocampus. Damage to the thalamus causes sensory problems and behavioral and cognitive changes. These challenges affect speech patterns, such as finding the right words for the proper response, difficulty connecting with others, and apathy affecting socialization and interaction.

Life experiences, including trauma, are learned and remembered for future reference. The hippocampus is the primary brain area for managing learning and memory processes. With coordinated efficiency, the hippocampus operates by integrating feedback from the amygdala (the fear circuit) and other brain structures via the thalamus to oversee learning and memory, playing a significant role in goal-directed responses to past traumas and threat-related emotions (Cozolino, 2010). The hippocampus is essential for evaluating sensory input from the amygdala and comparing the similarities and differences between past and present emotions to elicit a goal-directed response for survival.

Consequently, a decrease in the size of the hippocampus has been observed in some individuals with PTSD, where cell damage and shrinkage occur due to elevated cortisol levels released during trauma responses. This reduction in the hippocampus also contributes to misunderstanding and misinterpreting threat-related information, leaving survivors vulnerable to incorrect responses to stimuli (van der Kolk, 2014). The anterior cingulate cortex (ACC) is another crucial brain region that links cognitive functions and autonomic responses with other areas involved in intense emotions and cognitive tasks necessary for surviving traumatic experiences (Cozolino, 2021; LeDoux, 1996; Siegel, 2007). Known as the "chief operating officer," the ACC allocates attentional resources for survival and regulates emotional arousal (Siegel, 2007, p. 114). As part of the frontal lobe, the ACC's primary role is to activate working memory to perform cognitive tasks that require selective attention, resource allocation, decision-making, and control of physical responses to traumatic experiences (LeDoux, 1996). The ACC is often underactive in individuals with PTSD, and damage to this area results in apathy, depression, emotional disconnection, and disorientation (LeDoux, 1996; Sweeton, 2017).

As mentioned, the amygdala protects against threats by constantly assessing risks in a person's environment. Therefore, understanding its role is crucial when implementing trauma-informed, culturally based intervention (TICBI) (see Chaps. 7, 8, and 9). For example, when the amygdala becomes hyperresponsive, the size of the left hippocampus shrinks, leading to emotional arousal and re-experiencing trauma, known as PTSD flashbacks (Careaga et al., 2016; Desmedt et al., 2015; Michopoulos et al., 2016; Wicking et al., 2016). Recognizing the connections and disconnects between these areas helps explain why individuals exposed to life-threatening events in war, conflict, or chronic abuse often report feeling less safe and experience heightened perceived threats (Grieger et al., 2003; Ramsay & Woods, 2014). PTSD influences the actions and behaviors of survivors, resulting in constant vigilance, anxiety, and affecting their quality of life and well-being, as well as impacting their offspring's development (Heim & Nemeroff, 2001). These responses and variations among survivors can be observed in the regions indicated in Fig. 3.1, through functional magnetic resonance imaging (fMRI) and positron emission tomography (PET) scans of brain activity during trauma responses (LeDoux, 2002; Porges, 2007; van der Kolk, 2014).

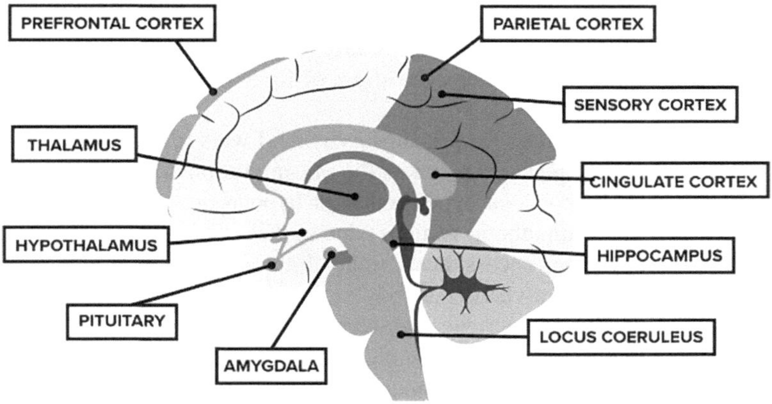

Fig. 3.1 Brain parts involved in a stress and fear response. Source: created by the trauma-informed, culturally based intervention (TICBI) team (Bokore, 2022)

5 The Long-Term Impact of Neurobiological Changes

The consequences of trauma persist across generations, potentially causing cellular alterations during gestation that affect the function of specific genes. This is particularly concerning for refugee children who may experience trauma transference, which begins in gestation and leads to what is termed an "epigenetic adjustment," or the alteration of the DNA sequence in the unborn children of refugees (van der Kolk, 2014, pp. 152–155). To prevent intergenerational trauma, understanding this gene modification is crucial for social workers providing TICBI. I have observed these changes in my community, impacting younger generations and inflicting generational pain on the community. Siegel (2011, p. 277), referencing Canadian researcher Dr. Michael Meaney's epigenetic research, explains how trauma transference occurs during gestation and in the early years of mother–child attachment. This transference continues for refugee children affected by inadequate nurture from traumatized or overstressed mothers, inhibiting the epigenome of specific genes and altering the child's experience-dependent brain. The accumulation of stressful experiences during forced migration heightens the vulnerability for these incidents to occur, allowing war-related trauma to persist in subsequent generations.

However, there is hope for recovery and healing as children receive in their lives what is known as "cross-fostering," or emotional surrogacy. In this process, harmful childhood experiences can be compensated for by the presence of at least one supportive and nurturing person within the child's life or environment (McCormack et al., 2011; Meaney, 2001; Meaney & Ferguson-Smith, 2010; Siegel, 2011; Yehuda et al., 2014). Receiving support reverses the transgenerational impact of trauma, becoming a healing anchor for children (McCormack et al., 2011; Meaney, 2001). These anchors can be family members, teachers, or social workers who act as emotional lifelines, counteracting negative childhood experiences (Matto et al., 2014; McCormack et al., 2011; Meaney, 2001; Meaney & Ferguson-Smith, 2010). Intervention starts by understanding the stress of forced migration, which happens in stages.

For example, it is common for female survivors of wars to experience the initial stage of forced migration (also known as the pre-migration stage), facing extreme stress characterized by constant hyperarousal or hypoarousal (the activation of the sympathetic and parasympathetic systems) due to the anticipation of war and environmental conditions that necessitate a fight, flight, or freeze response (Bokore, 2019). These emotions overwhelm the body's and brain's responses, impacting their well-being. The stressful experience continues in the second stage of migration, known as the flight or transit stage. This phase, which represents a more devastating part of forced migration, entails traumatic events en route to an interim or refugee-hosting country for safety, where their fight, flight, or freeze (hypoarousal in the parasympathetic nervous system) response persists. For many refugees, the third stage of migration or the post-settlement phase sees their stress responses continue as some struggle with the challenges of establishing social connections or integrating into the host country. For instance, some stories from participants in my TICBI project highlight the stress of adjusting to post-war life in a new country, adapting to a different culture, and facing systemic marginalization and oppression, which trigger memories of violence and deprivation. This is particularly challenging for survivors from non-white communities who encounter resettlement barriers that continue to create epigenetic challenges for generations.

6 The Impact of Epigenetic Challenges

East et al.'s (2018) study on Somali refugee mothers shows a direct correlation between mothers' traumatic experiences and their children's well-being, a suggestive pattern of intergenerational trauma. A mother's past trauma causes depressive moods and "volatility panic symptoms" (p. 279), affecting early attachment emotional development. Similarly, many mental health problems are a result of failed caregiving relationships in a child's early years, creating multiple social and emotional barriers in their adult life (East et al., 2018; Schore, 2001).

The children of refugee parents depend on their parents in their early years to develop their capacity to cope with environmental stressors in their adult life, allowing them to expand their motor and sensory abilities to interact with their social environment (Schore, 2012). Refugee parents, particularly those who experienced extreme trauma, are emotionally unable to attend to their child's emotional needs. Similarly, the toxic environment created by their parents' past/present experiences may lead to a disoriented adult life for the child that will perpetuate trauma transference into the next generation. Table 3.1 shows the trauma responses for Somali mothers (first generation) and youths (second generation, based on findings from my studies in 2014 and 2018.

Table 3.1 The impact of war-related trauma for first- and second-generation survivors

First-generation	Second-generation
A refugee parent living with an active sympathetic response and exhibiting hypervigilant behavior: • Preoccupied with the past and repeatedly telling stories of victimhood, tribal glory, survival, hate, revenge, and longing to return home • Living with daily stressful emotions, lack of helpful social connections, and being an emotionally absent parent, overlooking most of their children's needs • Has developed maladaptive coping skills that include preoccupation with social media and being involved in online fights • Over-involvement with political structures and division of power in the country of birth or resettlement countries • Has an inadequate understanding of the resettlement country's systems, including schooling and education, to support their children's growth and development	A child of a refugee parent becomes a hypervigilant adult with avoidance behaviors: • Dealing with PTSD and has difficulty describing childhood experiences and presents an incoherent family history (verbal memory deficits) • Has developed conflicted emotions that include wanting to get close to people, including family, at the same time showing avoidance behaviors and self-isolation • Avoids emotional responses and hides behind religious and community activities • Adapts parental fears/narratives of relationships including seeing the others outside the family, the religious community, or the tribe as dangerous • Struggles with educational underachievement, hopelessness, isolation; is part of group with a high level of school dropout • Lives in fear and is vulnerable to joining the wrong group for safety

Adapted from Bokore (2016a, 2016b, 2017, 2019) studies and publications

As seen in this table, trauma ominously impacts not only the first but also subsequent generations as well. For children of survivors, persistent parental stress interferes with their development, causing neurobiological changes that impact their emotional, social, and stress responses. The first generation's trauma response becomes an indicator of how trauma will shape the future of subsequent generations. Hence, first-generation responses are repeated by the second generation.

7 Epigenetic Challenges and Transgenerational Trauma

If the effects of trauma are passed down directly from the first generation to the next generations causing what is known as intergenerational trauma or transgenerational trauma is when the offspring of the first and second generations transfer their trauma responses to subsequent generations who have not been directly exposed to the trauma (Brave Heart et al., 2015). In the past few decades, transgenerational trauma has been studied in the context of African American slavery (Crawford et al., 2003), Holocaust survivors (Wiseman et al., 2006), and Indigenous people in the Americas (Brave Heart et al., 2015; Quinn et al., 2018).

Not all parental traumas result in intergenerational trauma (Schore, 2012, p. 386); while some children are affected by the quality of parent/child interaction or their childhood environments, others overcome adversities with the help of available support systems (Eisenberg, 1998, p. 213; Schore, 2012, p. 6). The long-term challenge for parent survivors is the perpetuation of a toxic environment, increasing the constant activation of their limbic system. This life prevents them from providing appropriate, emotionally charged, right-brain-to-right-brain communication with and connection to their children. These parental experiences of toxic stress affect their adult children's ability to cope with emotionally challenging events later in life (Cozolino, 2010; Porges, 2007; Porges & Dana, 2018; Siegel, 2011; van der Kolk, 2014). The good news for service providers and their clients is knowing that with adequate help and time, even as trauma is transferable, it is also reversible with the help of TICBI (see Chapters 7–10).

Definitions

Amygdala: A sensory processor of information determining the level of danger in the brain's subcortical area. It alerts areas of the brain involved in safety responses. Responses include behaviors that avoid painful memories or social activities. As part of the brain's limbic system, the amygdala is responsible for processing reactions related to fear and anxiety and forming emotional memories. It helps us recognize emotions in ourselves and others, influences decision-making, and contributes to the "fight or flight" response.

Brainstem: Located at the bottom of the head, the brainstem connects the brain and spinal cord. It includes the midbrain, pons, and medulla oblongata, and manages life-sustaining activities such as breathing, heart rate, and balance.

Cerebral cortex: The outer layer of the brain, covering the cerebrum. It manages essential human functions such as memory, thinking, learning, solving problems, reasoning, emotions, consciousness, and unconscious sensory functions.

Hippocampus: This is the brain's timekeeper and memory storage. With each experience, this area is activated to remind humans about the similarities and differences between current and past experiences. A decrease in the hippocampus's volume notably affects memory and the ability to learn new things.

Hyperarousal: An internal anxiety and emotional reaction leading to irritability, anger, and a heightened reaction to being startled, affecting a survivor's ability to concentrate or have a good night's sleep.

Hypervigilance: Experiencing constant threat and forcing the assessment of potential dangers. Behavior that creates difficulty connecting or socializing with others, increased anxiety, and lack of sleep and fatigue.

Hypoarousal is a parasympathetic nervous system response that activates a freeze response (a dorsal vagal response), shutting off the sympathetic nervous system as the person feels escape is impossible.

Hypothalamus: The hypothalamus is a small part of the brain that controls many bodily functions such as hormone production, regulating body temperature and the sleep cycle. It also controls emotions, behavior, and memory.

Interoception: Sensation of our internal body signals.

Limbic system: A group of brain structures significantly regulates emotions, memory, behavior, and motivation. It connects the functions of the cerebral cortex (for higher-level thinking) and the brainstem (for basic bodily functions). Its additional roles include processing emotions, forming memories, and prompting survival behaviors (eating, reproduction, and the fight-or-flight response or danger alert).

Medial prefrontal cortex (mPFC): A brain region in the frontal lobe. It is responsible for decision-making, mindful awareness, attention, judgment, and consolidative activities, making the amygdala reaction less active.

Midbrain: Also called the mesencephalon, the midbrain is part of the central nervous system. It is a relay system that shares the information necessary for vision, hearing, and other functions, including motor movement, pain sensation, and the sleep/wake cycle.

Neuroception: Environmental detection without awareness. How the autonomic nervous system interacts with the world around us for safety.

Noradrenergic system: The neuronal system synthesizes, stores, and releases the norepinephrine neurotransmitter. The locus coeruleus (LC) contains norepinephrine (NE), which is synthesized by neurons. Therefore, the LC-NE system plays a significant role in arousal, attention, and stress response.

Parasympathetic nervous system (PNS): Responsible for calming the body after a stressful or fearful experience. PTSD is often caused by the inability of the nervous system to discharge the energy created by the SNS after the threat has passed.

Prefrontal cortex: Manages higher-level cognitive functions, including planning, decision-making, working memory, personality, social behavior, and emotional regulation. It directs other brain areas in cognitive function and impulse control. The area is fully developed during adulthood, impacting information processing, increasing the vulnerability for children and young when there are life-threatening situations.

Sympathetic nervous system (SNS): The brain area activated for survival during trauma by increasing the production of cortisol and adrenaline (hormones) for escape. Consistent hyperactivity of the sympathetic branch of the autonomic system is expected for survivors dealing with PTSD.

Discussion Questions

1. How do the ICD-11 and DSM-5 define PTSD? Explain the differences, including when and why you would use each.
2. Explain the main differences between civilian and veteran experiences of war-related trauma.
3. What will you consider when mapping your client's migration experiences that impact their mental health and create resettlement and integration challenges?
4. Describe your understanding of the brain functions involved in experiencing trauma and parent–child relationships. How can this knowledge assist you in client assessment, psychoeducation, and preventing intergenerational and transgenerational trauma?
5. When working with adults who grew up in toxic environments or with parents who survived prolonged trauma, why is it important to examine trauma from a neurodevelopmental perspective?
6. Why should social workers care about brain processes and trauma responses?

7. Define and/or provide examples:
   - The causes of refugee trauma and neurobiological responses.
   - Primary and secondary trauma: Describe how it manifests in refugees.
   - Symptoms of PTSD in refugees:
   - Trauma transference and practice-based transference or countertransference.

Competing Interests The author has no conflicts of interest to declare that are relevant to the content of this chapter.

References

Ben-Ezra, M., Hyland, P., Karatzias, T., Maercker, A., Hamama-Raz, Y., Lavenda, O., Mahat-Shamir, M., & Shevlin, M. (2020). A cross-country psychiatric screening of ICD-11 disorders specifically associated with stress in Kenya, Nigeria and Ghana. *European Journal of Psychotraumatology, 11*(1).

Blume, A. W., Schmaling, K. B., & Marlatt, G. A. (2000). Revisiting the self-medication hypothesis from a behavioral perspective. *Cognitive and Behavioral Practice, 7*(4), 379–384. https://doi.org/10.1016/s1077-7229(00)80048-6

Bokore, N. (2016a). "*Somali-Canadian women: Historical past of survival and facing everyday challenges of resettlement". A dissertation submitted to the Faculty of Graduate Studies in partial fulfillment of the requirements for the degree of doctor of philosophy*. Graduate Program in Social Work, York University. https://yorkspace.library.yorku.ca/server/api/core/bitstreams/a978e567-6711-49d9-b1cd-e89ac81f68a4/content

Bokore, N. (2016b). Documenting refugee stories: Memories of survivors of East African war. *International Social Work Journal, 3*(1).

Bokore, N. (2017). Historical trauma, resettlement, and intervention strategies: An analysis of Somali-Canadians' experiences. *International Migration, 56*(2), 146–162. https://doi.org/10.1111/imig.12418

Bokore, N. (2019). Life history research method: Learning from the lives of female survivors of wars. In *Sage research methods Cases Part 2*. Sage. https://doi.org/10.4135/9781526490346

Bokore, N. (2022, August 1–5). Re-think resettlement and integration: Creating culturally based trauma-informed interventions. In *19th conference of the International Association for the Study of Forced Migration (IASFM), hosted by the Catholic University of Santos, Brazil.*

Bokore, N., McGrath, S. L., McGuire, P., Karod, A., Rahimpour, M., & Asokumar, A. (2023). Developing a trauma-informed culturally-based intervention (TICBI) approach for refugee resettlement practices. *International Journal of Social Work, 10*(1), 1. https://doi.org/10.5296/ijsw.v10i1.20561

Brave Heart, M. Y. H. B., Chase, J., Elkins, J., & Altschul, D. B. (2015). Historical trauma among indigenous peoples of the Americas: Concepts, research, and clinical considerations. *Journal of Psychoactive Drugs, 43*(4), 282–290. https://doi.org/10.1080/02791072.2011.628913

Brewin, C. R., Gregory, J. D., Lipton, M., & Burgess, N. (2010). Intrusive images in psychological disorders: Characteristics, neural mechanisms, and treatment implications. *Psychological Review, 117*(1), 210–232. https://doi.org/10.1037/a0018113.

Careaga, M. B. L., Girardi, C. E. N., & Suchecki, D. (2016). Understanding posttraumatic stress disorder through fear conditioning, extinction and reconsolidation. *Neuroscience & Biobehavioral Reviews, 71*, 48–57. https://doi.org/10.1016/j.neubiorev.2016.08.023

Cozolino, L. J. (2010). *The neuroscience of psychotherapy: Healing the social brain* (2nd ed.). W. W. Norton & Company.

Cozolino, L. (2021). *The development of a therapist: Healing others - healing self.* W. W. Norton & Company.
Crawford, J., Nobles, W., & Leary, J. (2003). Repatriations and healthcare for African Americans: Repairing the damage from the legacy of slavery. In R. Winbush (Ed.), *Should America pay: Slavery and the raging debate on repatriations.* HarperCollins Publishing.
Dana, D. (2020). *Polyvagal exercises for therapists and clients: 50 Client-Centered Practices.* W. W. Norton & Co.
De Jong, J. T., Komproe, I. H., & Van Ommeren, M. (2003). Common mental disorders in postconflict settings. *Lancet, 361*(9375), 2128–2130. https://doi.org/10.1016/s0140-6736(03)13692-6
Desmedt, A., Marighetto, A., & Piazza, P. (2015). Abnormal fear memory as a model for posttraumatic stress disorder. *Biological Psychiatry, 78*(5), 290–297. https://doi.org/10.1016/j.biopsych.2015.06.017
East, P. L., Gahagan, S., & Al-Delaimy, W. K. (2018). The impact of refugee mothers' trauma, posttraumatic stress, and depression on their children's adjustment. *Journal of Immigrant and Minority Health, 20*(2), 271–282. https://doi.org/10.1007/s10903-017-0624-2
Eisenberg, L. (1998). Nature, niche, and nurture. *Academic Psychiatry, 22*(4), 213–222. https://doi.org/10.1007/bf03340021
Evans, A., & Coccoma, P. (2014). *Trauma-informed care: How neuroscience influences practice.* Routledge.
Fillenz, M. (1990). *Noradrenergic neurons.* Cambridge University Press.
Grieger, T. A., Fullerton, C. S., Ursano, R. J., & Reeves, J. J. (2003). Acute stress disorder, alcohol use, and perception of safety among hospital staff after the sniper attacks. *Psychiatric Services, 54*(10), 1383–1387. https://doi.org/10.1176/appi.ps.54.10.1383.
Hansen, N. (2021). Locus coeruleus malfunction is linked to psychopathology in prodromal dementia with lewy bodies. *Frontiers in Aging Neuroscience, 13.* https://doi.org/10.3389/fnagi.2021.641101
Heim, C., & Nemeroff, C. B. (2001). The role of childhood trauma in the neurobiology of mood and anxiety disorders: Preclinical and clinical studies. *Biological Psychiatry, 49*(12), 1023–1039. https://doi.org/10.1016/s0006-3223(01)01157-x
Hinton, D. E., & Good, B. J. (2016). *Culture and PTSD: Trauma in global and historical perspective (The ethnography of political violence).* University of Pennsylvania Press.
Jones, E., Vermaas, R. H., McCartney, H., Beech, C., Palmer, I., Hyams, K., & Wessely, S. (2003). Flashbacks and post-traumatic stress disorder: The genesis of a 20th-century diagnosis. *British Journal of Psychiatry, 182*(2), 158–163. https://doi.org/10.1192/bjp.182.2.158
Kessler, R. C., Aguilar-Gaxiola, S., Alonso, J., Benjet, C., Bromet, E. J., Cardoso, G., Degenhardt, L., De Girolamo, G., Dinolova, R. V., Ferry, F., Florescu, S., Gureje, O., Haro, J. M., Huang, Y., Karam, E. G., Kawakami, N., Lee, S., Lepine, J., Levinson, D., et al. (2017). Trauma and PTSD in the WHO world mental health surveys. *European Journal of Psychotraumatology, 8*(sup5). https://doi.org/10.1080/20008198.2017.1353383
Kienzler, H. (2008). Debating war-trauma and post-traumatic stress disorder (PTSD) in an interdisciplinary arena. *Social Science & Medicine, 67*(2), 218–227. https://doi.org/10.1016/j.socscimed.2008.03.030
Kleinman, A., & Kleinman, J. (1991). Suffering and its professional transformation: toward an ethnography of interpersonal experience. *Culture, Medicine and Psychiatry, 15*(3), 275. https://doi.org/10.1007/bf00046540
Knight, C. (2015). Trauma-informed social work practice: Practice considerations and challenges. *Clinical Social Work Journal, 43*(1), 25–37. https://doi.org/10.1007/s10615-014-0481-6
Lanius, R., & Olff, M. (2017). The neurobiology of PTSD. *European Journal of Psychotraumatology, 8*(1). https://doi.org/10.1080/20008198.2017.1314165
LeDoux, J. E. (1995). Emotion: Clues from the Brain. *Annual Review of Psychology, 46*(1), 209–235. https://doi.org/10.1146/annurev.ps.46.020195.001233.

LeDoux, J. (1996). *The emotional brain: The mysterious underpinnings of emotional life*. Simon and Schuster.
LeDoux, J. E. (2002). *Synaptic self: How our brains become who we are*. Viking Adult.
Levine, P. (2015). *Trauma and memory: Brain and body in a search for the living past: A practical guide for understanding and working with traumatic memory*. North Atlantic Books.
Matto, H. C., Strolin-Goltzman, J., PhD, & Ballan, M. S. (2014). *Neuroscience for social work: Current research and practice*. Springer Publishing Company.
McCauley, J. L., Killeen, T., Gros, D. F., Brady, K. T., & Back, S. E. (2012). Posttraumatic stress disorder and co-occurring substance use disorders: Advances in assessment and treatment. *Clinical Psychology, 19*(3), 283–304. https://doi.org/10.1111/cpsp.12006.
McCormack, L., Hagger, M. S., & Joseph, S. (2011). Vicarious Growth in Wives of Vietnam Veterans: A Phenomenological Investigation Into Decades of "Lived" Experience. *The Journal of Humanistic Psychology, 51*(3) 273–290. https://doi.org/10.1177/0022167810377506.
McLean, C. P., Asnaani, A., & Foa, E. B. (2015). Prolonged exposure therapy. In U. Schnyder & M. Cloitre (Eds.), *Evidence based treatments for trauma-related psychological disorders: A practical guide for clinicians* (pp. 143–159). Springer International Publishing/Springer Nature. https://doi.org/10.1007/978-3-319-07109-1_8.
Meaney, M. J. (2001). Maternal care, gene expression, and the transmission of individual differences in stress reactivity across generations. *Annual Review of Neuroscience, 24*(1), 1161–1192. https://doi.org/10.1146/annurev.neuro.24.1.1161.
Meaney, M. J., & Ferguson-Smith, A. C. (2010). Epigenetic regulation of the neural transcriptome: the meaning of the marks. *Nature Neuroscience, 13*(11), 1313–1318. https://doi.org/10.1038/nn1110-1313.
Michopoulos, V., Vester, A., & Neigh, G. (2016). Posttraumatic stress disorder: A metabolic disorder in disguise? *Experimental Neurology, 284,* 220–229. https://doi.org/10.1016/j.expneurol.2016.05.038.
Naegeli, C., Zeffiro, T., Piccirelli, M., Jaillard, A., Weilenmann, A., Hassanpour, K., Schick, M., Rufer, M., Orr, S. P., & Mueller-Pfeiffer, C. (2018). Locus coeruleus activity mediates hyperresponsiveness in posttraumatic stress disorder. *Biological Psychiatry, 83*(3), 254–262. https://doi.org/10.1016/j.biopsych.2017.08.021.
Ogden, P., & Fisher, J. (2015). *Sensorimotor psychotherapy: Interventions for trauma and attachment (Csm)*. W. W. Norton & Company.
Ogden, P., Minton, K., & Pain, C. (2006). *Trauma and the body: A sensorimotor approach to psychotherapy* (Norton series on interpersonal neurobiology) (1st ed.). W. W. Norton & Company.
Ona, P. E. Z. (2020). *Acceptance and commitment therapy for borderline personality disorder: A flexible treatment plan for clients with emotional dysregulation*. Context Press.
Porges, S. W. (2007). The polyvagal perspective. *Biological Psychology, 74*(2), 116–143. https://doi.org/10.1016/j.biopsycho.2006.06.009
Porges, S. W., & Dana, D. (2018). *Clinical applications of the polyvagal theory: The emergence of Polyvagal-Informed therapies (norton series on interpersonal neurobiology)*. W. W. Norton & Company.
Quinn, M. G., Hokanson, L., Schüz, N., De Salas, K., & Scott, J. (2018). A systematic review of indigenous caregiver functioning and interventions. *Quality of Life Research, 27*(8), 2007–2017. https://doi.org/10.1007/s11136-018-1836-1
Ramsay, D. S., & Woods, S. C. (2014). Clarifying the Roles of Homeostasis and Allostasis in Physiological Regulation. *Psychological Review, 121*(2), 225–247. https://doi.org/10.1037/a0035942.
Sareen, J. (2014). Posttraumatic stress disorder in Adults: impact, comorbidity, risk factors, and treatment. *The Canadian Journal of Psychiatry, 59*(9), 460–467. https://doi.org/10.1177/070674371405900902
Schaffer, J. B., & Rodolfa, E. (Eds.). (2017). *An ICD-10-CM casebook and workbook for students: Psychological and behavioral conditions*. American Psychological Association.

Schore, A. N. (2001). Effects of a secure attachment relationship on right brain development, affect regulation, and infant mental health. *Infant Mental Health Journal, 22*(1–2), 7–66. https://doi.org/10.1002/1097-0355(200101/04)22:17::AID-IMHJ23.0.CO;2-N

Schore, A. N. (2012). *The science of the art of psychotherapy: The latest work from a pioneer in the study of the development* (1st ed.). W. W. Norton & Company.

Siegel, D. J. (2007). *The mindful brain: reflection and attunement in the cultivation of well-being*. Norton & Company.

Siegel, D. J. (2011). Mindsight: *The New Science of Personal Transformation*. Bantam.

Simich, L. (n.d.). Refugees, forced migration & mental health: Recovery from trauma and the destigmatization of mental illness. In *Annual Educational Conference, Toronto, Canada. unpublished Metropolis Conference Toronto*.

Sweeton, J. (2017). *Trauma treatment toolbox: 165 brain-changing tips, tools & handouts to move therapy forward*. PESI Publishing.

Taché, Y., & Brunnhuber, S. (2008). From Hans Selye's discovery of biological stress to the identification of corticotropin- releasing factor signaling pathways. *Annals of the New York Academy of Sciences, 1148*(1), 29–41. https://doi.org/10.1196/annals.1410.007.

Tedeschi, R., & Calhoun, L. G. (1995). *Trauma and transformation: Growing in the aftermath of suffering*. SAGE Publications, Inc.

Tedeschi, R. G., & Calhoun, L. G. (1996). The posttraumatic growth inventory: Measuring the positive legacy of trauma. *Journal of Traumatic Stress, 9*(3), 455–471. https://doi.org/10.1002/jts.2490090305.

Van Der Kolk, B. A. (2014). *The body keeps the score: Brain, mind, and body in the healing of trauma*. Penguin Publishing Group.

Volpicelli, J., Balaraman, G., Hahn, J., Wallace, H., & Bux, D. (1999). The role of uncontrollable trauma in the development of PTSD and alcohol addiction. *Alcohol Research and Health, 23*(4), 256–262. https://pubmed.ncbi.nlm.nih.gov/10890822

Wicking, M., Steiger, F., Nees, F., Diener, S. J., Grimm, O., Ruttorf, M., Schad, L. R., Winkelmann, T., Wirtz, G., & Flor, H. (2016). Deficient fear extinction memory in posttraumatic stress disorder. *Neurobiology of Learning and Memory, 136*, 116–126. https://doi.org/10.1016/j.nlm.2016.09.016

Wilkinson, M. (2010). *Changing minds in therapy: Emotion, attachment, trauma, and neurobiology*. W. W. Norton & Company.

Wiseman, H., Metzl, E., & Barber, J. P. (2006). Anger, guilt, and intergenerational communication of trauma in the interpersonal narratives of second generation holocaust survivors. *American Journal of Orthopsychiatry, 76*(2), 176–184. https://doi.org/10.1037/0002-9432.76.2.176

World Health Organization. (2020). World health statistics 2020: Monitoring health for the SDGs, *Sustainable Development Goals*. World Health Organization.

Yehuda, R., Daskalakis, N. P., Lehrner, A., Desarnaud, F., Bader, H. N., Makotkine, I., Flory, J. D., Bierer, L. M., & Meaney, M. J. (2014). Influences of maternal and paternal PTSD on epigenetic regulation of the glucocorticoid receptor gene in holocaust survivor offspring. *The American Journal of Psychiatry, 171*(8), 872–880. https://doi.org/10.1176/appi.ajp.2014.13121571

Chapter 4
The Development of Post-Traumatic Growth

1 What Is Post-Traumatic Growth?

Post-traumatic growth is the positive change that follows a traumatic experience (Fujisawa et al., 2015; Mattson et al., 2018; Nakagawa et al., 2016; Tedeschi & Calhoun, 1995, 2004). Many factors are involved in the development of post-traumatic growth. For example, a study by Ochsner et al. (2002, p. 1215) revealed the correlation between the development of post-traumatic growth and a survivor's cognitive reappraisal. Based on functional magnetic resonance imaging (fMRI), the study evaluated the brain activity of 15 healthy adults and their responses to negative images, asking them to rate what they saw as "increasing, maintaining, or decreasing" their emotional responses. Their responses are explained as the results of their "cognitive reappraisal" (p. 1224), which changes both subjective experience and emotional reactions. Survivors develop PTSD when their cognitive reappraisal activates their lateral and medial prefrontal cortex (PFC), decreasing activities in the area of the medial orbitofrontal cortex and amygdala. The cognitive reappraisal modulates the thoughts and feelings associated with traumatic experiences.

The functions of the prefrontal cortex and cognitive reappraisal helped me manage my emotions as a young mother of four. They allowed me to pay attention to my children's feelings and seek resources to change my family's circumstances. Years later, working with survivors taught me the importance of this capacity for them and the positive impact it has on developing post-traumatic growth (PTG). It is the essence of my resilience, providing me strength throughout my forced migration and helping me protect my children from the effects of their traumatic past.

Tedeschi and Calhoun (1995) first researched and wrote about this process and its benefits for promoting positive change after trauma. This work shifted the discussion from the impacts of PTSD or other mental illness to recovery and personal growth (Abraham et al., 2018; Ai et al., 2007; Simsir & Dilmac, 2018; Tedeschi et al., 2018; Uy & Okubo, 2018). The growing recognition of post-traumatic growth

N. Bokore, *Trauma-Informed, Culturally Based Intervention*,
https://doi.org/10.1007/978-3-032-06835-4_4

as part of the post-traumatic stress disorder (PTSD) led to a paradigm shift in trauma studies, focusing less on the adverse effects of PTSD and more on what fosters positive change in a survivor's life, increasing not only their capacity for survival but also promoting well-being.

Post-traumatic growth is based on hope and survival, which stems from positive interoception, self-regulation, and self-perception. It helps survivors overcome the adversities of forced migration and resettlement, increasing their sense of safety. For some survivors, distinguishing environmental stimuli is part of their physiological/emotional capacity. Reacting or responding to stimuli comes from the primitive part of the brain that manages the implicit survival instinct and functions without conscious awareness. It is what makes some survivors vulnerable to developing PTSD (see Chap. 3) while others develop post-traumatic growth (Shaw et al., 2005; Tedeschi et al., 2018). The exciting part of the development of post-traumatic growth is how survivors are able not only to survive the traumatic event but also reach beyond what they would have been able to achieve before their traumatic experiences (Nes & Segerstrom, 2006; Prati & Pietrantoni, 2009; Tedeschi & Calhoun, 2004). (This is worth noting when providing TICBI, especially when working with those with an active, engaged mindset [see Chaps. 8 and 9].) A correlation exists between assessing internal feelings (positive interoception) and responding to environmental alerts (neuroception). Having a positive interoception helps with adaptive responses to environmental stressors.

2 Factors Contributing to Post-Traumatic Growth

2.1 Self-Perception

Daryl Bem (1967) introduced self-perception theory as an alternative to cognitive reappraisal theory. This theory explains how humans adjust their emotions and attitudes during challenging times (e.g., forced migration) and change their decision-making based on interpersonal processes and observations. Internal capacity changes the process of cognitive reappraisal and self-perception. It changes a survivor's reaction to stressors and allows them to seek connection to increase their sense of survival. Self-perception minimizes the adverse effects of trauma, including the development of PTSD. Religious beliefs and cultural values also contribute to a person's self-perception, providing internal strength and resilience.

2.2 The Role of Supportive Resources

As discussed in Chapter One, refugees are welcomed and accepted based on ethnicity, religion, and cultural similarity to the host nation's population. These variations give those accepted as part of the community/society a sense of entitlement, confidently asking for services and resources contributing to their PTG. At the same time, those who are excluded experience guilt and sometimes blame themselves to rationalize their lack of access to resources, leading to PTSD-related symptoms. I remember arriving in Italy as an Ethiopian refugee, and how public support was readily available for me due to the historical ties. That available support made a difference for me and my family while we waited for resettlement in a third country. However, when I arrived in Canada, my feelings and experiences changed because of an initial lack of access to services and the microaggression I experienced as a Black person, triggering my past trauma, shrinking my window of tolerance. There were days when I experienced and lived through the reaction and action of my hyperarousal (fight-or-flight mode) and, at times, hypoarousal (loss of hope) activation, which impeded my resettlement for the first decade after my arrival. It took a while, but my cultural values, faith, and internal resilience eventually kicked in. I began connecting with supportive individuals and agencies who provided me with needed resources. The combination of my changed attitude, self-perception, and self-concept further re-ignited my strength and belief in hard work, which are part of my philosophy of life—my post-traumatic growth response that helped me rebuild a life for my family.

2.3 Trauma and Positive Outlook

In their extensive research on post-traumatic growth, Richard Tedeschi and Lawrence Calhoun (1995, 1996, 2014) explain the role of culture and early life nurture in building a positive self-perception and outlook on life. For me, the supportive friends I met in my journey, religion, and culture provided me with positive feelings and a philosophy of life that helped me make meaning of events as they happened, increasing my positive attitude and hope for the future. It is an outlook that decreased my stress, providing me with strength and a sense of meaning to life that helped me overcome resettlement barriers. The human brain is neurobiologically structured to overcome adverse events, making trauma recovery possible and trauma-informed, culturally based intervention (TICBI) effective. Health psychologists link the ability to overcome adversities with internal resilience and external resources and support systems that contribute to the strength-building process of human growth and healing (Hadad, 2013; Tedeschi et al., 2018).

3 The Neurobiology of Post-Traumatic Growth

Post-traumatic growth (PTG) neurobiology involves changes in brain structure and function, including strengthening connections in areas associated with positive emotions, social bonding, and resilience. This strengthening is linked to internal ability and the availability of external resources, activating the survivor's prefrontal cortex to identify triggers (safety or danger) and the survivor's meaning-making process. This unique ability of the human brain for recovery from trauma includes using past and present information to understand whether a trigger is part of a safe memory, causing calm (parasympathetic activation), or danger, needing sympathetic activation to escape. Humans are fundamentally social beings, preferring to seek connections with others, and the availability of support systems makes a difference in the assessment/evaluation of a situation and the meaning-making process. Trauma severs these connections, causing PTSD (Porges, 2007; van der Kolk, 2014). Some survivors can recognize the availability of support systems, signaling positive connection and emotional co-regulation.

This ability provides survivors with a sense of safety and increases their motivation to seek resources, aiding in the development of post-traumatic growth. As mentioned above, for some survivors of wars, social support is lacking, creating a barrier to the growth they might achieve. Although post-traumatic growth is understood and has positive outcomes for survivors, the neurobiological structures and brain processes linking the social brain to post-traumatic growth are not yet well-researched or fully understood. Louise Cozolino (2010, p. 179) explains that the neurobiologically activated need for social connection is evident in the interdependency of neurons and the human tendency to be social beings. With adequate support, survivors can feel safe enough to engage in establishing social connections, which helps them manage PTSD triggers and benefit from post-traumatic growth.

However, the development and maintenance of post-traumatic growth is linked to what Cozolino (2010, p. 179) calls the role of the "social synapse," which helps survivors grow when they are affected by challenging environments. This highlights the evolutionary brain's limitless capacity, making PTG possible (Cozolino, 2010; LeDoux, 1996; Siegel, 2007; Tedeschi & Calhoun, 1995, 1996). Throughout its evolutionary history, the human brain has expanded from relying primarily on the functions of the brain stem and limbic system, two original areas wired to handle survival instincts that assess and respond to emotional and relational experiences. The evolved, newer brain continues to grow, particularly in the regions involved in social care, survival, reasoning, problem-solving, planning, analyzing, and interpreting internal sensations and environmental activities (Cozolino 2010; Levine, 2015; Siegel, 2011). When survivors experience post-traumatic growth, their social synapse enables them to analyze traumatic memories or triggers and use facts to foster a positive response, activating their ventral vagal system and enhancing their biological need to connect (Porges, 2007). It is a natural process that creates neurobiological changes that reduce stress responses, increase positive social engagement, promote resilience, and contribute to post-traumatic growth (Abraham et al., 2018; Saleebey, 2013; Tedeschi & Calhoun, 1995).

Social support and cultural adaptation are at the core of post-traumatic growth, shaping the neurobiological changes that affect executive function and the emotional regulation area of the brain during and after traumatic events. When providing TICBI, it is vital to recognize that post-traumatic growth is distinct from resilience. While resilience pertains to internal biological or genetic processes that enable positive post-trauma change, post-traumatic growth, as previously mentioned, refers to the "growth" that occurs during the post-trauma phase, calming limbic responses and enhancing opportunities for "new ways of thinking, feeling, and behaving" after trauma (Tedeschi et al., 2018, p. 5).

In the past two decades, personal resilience has increasingly received attention in trauma intervention (Abraham et al., 2018; Tedeschi & Calhoun, 1995, 1996). In my personal and professional life, I have learned that although resilience is supported by an internal predisposition (genetic), the support of external systems is necessary for survivors to flourish. With both internal and external resources fostering neurobiological changes, survivors with post-traumatic growth benefit from their self-efficacy, self-confidence, and self-worth, enhancing their chances of recovery (Abraham et al., 2018; Blechman, 2000; Tedeschi & Calhoun, 1995, 1996).

3.1 *Subjective Responses to Environmental Cues*

Subjective responses to trauma are common among survivors of the Holocaust, the Rwandan genocide, the Vietnam War, and the civil wars in Somalia and Syria. Each survivor's interpretation and emotional response to their trauma ultimately defines their autonomic response system. While some may experience constant sympathetic nervous system (SNS) responses and remain alert to danger (active hyperarousal), others may perceive their circumstances as opportunities for change, utilizing available connections and developing resources for post-traumatic growth (PTG).

These changes are evident in survivors' responses, as the autonomic systems of traumatized individuals react to environmental cues and operate hierarchically, triggering the survivor's body–brain response (high or low arousal in response to a threat or calmness and readiness to socialize). These outcomes depend on subjective reactions. For example, those who develop PTSD have an active sympathetic response system that alerts their body and brain to the sensation of a threat by moving their nervous system into high alert or a heightened arousal state (increased heart rate, rapid breathing, tense muscles, and elevated stress hormones). In comparison, those who experience post-traumatic growth evaluate similar situations calmly, producing a parasympathetic response that maintains normal arousal levels, reduces heart rate, and regulates breath, leading to a state of relaxation and enhanced cognitive processing.

These subjective responses help them decrease stress and hormone production, triggering a ventral vagal response that fosters socialization and offers opportunities to benefit from support resources (see Porges, 2007, 2017; Porges & Dana, 2018). It is crucial to note that individuals with post-traumatic growth may also occasionally respond to their implicit survival reactions (see Chap. 9, Active, Engaged Mindset), which can overwhelm their emotions and evoke traumatic memories, temporarily shutting down their prefrontal cortex, which would otherwise remain active. During these moments, survivors need access to support systems to manage their stress (see Chap. 10).

3.2 The Danger of Toxic Stress to Post-Traumatic Growth

Occasional stress is part of life; it happens to everyone. Typical daily stress helps with productivity and contributes to a positive sense of self for those with post-traumatic growth. However, survivors often resettle in areas where toxic or extreme amounts of stress disrupt their usual post-traumatic growth response (see Chap. 9, Active, Engaged Mindset). Neuroimaging studies have shown the adverse effects of toxic stress on the activities of the prefrontal cortex, especially the dorsolateral prefrontal cortex (DLPFC), which promotes resilience, emotional and social intelligence, and post-traumatic growth. It also functions as a cognitive control center, allowing individuals to manage emotions, make complex social decisions, and recover from adversity (Cozolino, 2010).

Living in an environment of toxic stress also decreases the activation of the prefrontal cortex (PFC) and dorsal anterior cingulate cortex (ACC), which are linked to emotional strength and stability. Unfortunately, many resettled survivors of wars experience toxic stress due to economic precarity. Survivors often live in toxic environments (the presence of urban violence, lack of safety, and extreme surveillance by the police), which plays a vital role in the cognitive assessment of a stressful situation (neuroception). This increases their maladaptive behaviors (Fujisawa et al., 2015; Mattson et al., 2018; Nakagawa et al., 2016).

According to Porges (2007), the autonomic nervous system subconsciously evaluates environmental changes that may affect well-being, including those caused by toxic stress and threats. Survivors respond instinctively to environmental cues, including those in social settings, which is related to neuroception (van der Kolk, 2014, p. 129). They constantly evaluate their ecological situation positively, triggering an autonomic response, which motivates some survivors to look for growth opportunities beyond survival (Siegel, 2011, p. 370).

4 Understanding Post-Traumatic Growth in Social Work Practice

One of the social worker's roles in providing trauma-informed, culturally based intervention (TICBI) for survivors living in toxic environments is to become a reliable support for them and possess the skills and ability to help them regulate their emotions. This includes offering a safe space that promotes calm (parasympathetic response) and co-regulation. Mirror neurons play a significant role in co-regulation within the therapeutic interaction (see Chap. 10), centered on the emotional exchange between the social worker and the client, encouraging a responsive interaction to help each other manage emotions and activate the nervous system, creating a connection. Daniel Siegel (2010, pp. 335–336) refers to this connection and function within therapeutic settings as "emotional resonance," which describes a mutual understanding between the social worker and the client to foster harmony, calm, and productive interaction within the therapeutic relationship.

The human autonomic nervous system is known to begin developing a structure for relationships in early childhood (Fosha et al., 2009; Schore, 2002; Siegel, 2011). It is a structure that significantly impacts child development, defining healthy attachment, as well as the transference of trauma, depending on the quality of relationships and connections the child has with caregivers. These factors determine whether a child later becomes an adult who tends toward an active ventral vagal state (favorable for socialization), sympathetic activation (fight or flight response), or dorsal vagal activation (emotional shutdown). Being emotionally regulated early in life also enhances a child's ability later in life to develop resilience and growth during challenging times. For social workers, understanding child development is crucial in trauma therapy to recognize how early life experiences determine trauma responses.

The identification experiences starting with early life help social workers create an atmosphere of safety, trust, and the possibility of benefiting from co-regulation during therapeutic interactions, which is essential when providing TICBI. This approach contributes to post-trauma growth and resilience, drawing from the two primary sources of healing found in the wisdom and religious teachings of ancient Hebrews, Greeks, Christians, Buddhists, Hindus, and Muslims, rooted in understanding and with compassion to help survivors, particularly those grappling with shame and self-criticism to work toward recovery and healing. For many ethnic groups, ancient wisdom and spiritual teachings help with co-regulation, increasing survivors' capacity to overcome adversity, fostering their resilience and growth. Therefore, when providing TICBI, this therapeutic approach is a tool to explore clients' experiences, helping survivors transcend their immediate crises and, for some, transform adversity into strength while collaborating with them and continuing to share resources for growth. Utilizing wisdom and spiritual teachings in TICBI helps to overcome early childhood experiences, celebrate survival, appreciate life, and seek new possibilities despite traumatic pasts (Hadad, 2013).

In disciplines focusing on interventions, such as social work, it should not be a surprise that religion, spirituality, faith, resiliency, and strength are positively associated with the development of post-traumatic growth (Hadad, 2013; Saleebey, 2013; Shaw et al., 2005; Tedeschi et al., 2018). In addition to the pre-existing survival system (genetics), faith and social support can be an external source contributing to post-traumatic growth, providing survivors with the strength to make positive changes (Shaw et al., 2005). Traumatic experiences disrupt internal homeostasis and the survivor's ability to achieve stability due to the presence of stress and an active stress response, leading to long-term physiological and psychological changes that affect well-being. Religion, faith, and culture are part of the regulating system that provides emotional balance and calmness, essential for feeling safe, lowering stress levels, and providing homeostasis.

Therefore, understanding the role of pre-existing strength, resilience, and external support systems (regulators) catalyzes positive change and transforms survivors' lives. Social work is a discipline that focuses on survivors recognizing their inner strength and the capacity to develop or maintain post-traumatic growth. TICBI relies on the recognition of the built-in human capacity for survival. Using a recovery-based approach, social workers become supportive agents promoting post-traumatic growth, helping survivors recognize their built-in survival traits that they used in the past to face "life's challenges and walk away" (Saleebey, 2013, p. 18).

5 Psychoeducation: For Re-igniting, Resiliency, and Promoting Post-Traumatic Growth

As discussed in Chaps. 1 and 2, survivors of wars and conflicts are repeatedly exposed to violence during forced migration while engaging their survival systems (hyperarousal and hypoarousal states) and abandoning socialization (disconnecting from their ventral vagal response system). For those in the resettlement stage, experiences of microaggressions from host societies, which can be intentional, unintentional, or even well-meaning, keep them in survival mode and contribute to their daily stress. To confront these emotions and foster post-traumatic growth, psychoeducation assists survivors in re-igniting neural connections, internal strength, and resilience.

In TICBI, psychoeducation is information sharing about the issue that brought your client to therapy and the varied human responses to it. I used to share information verbally or provide handouts that clients could use to guide their healing. I sometimes encouraged survivors to discover the needed information and practice their skills for adaptation and long-term change. Psychoeducation is about helping survivors self-understand and re-ignite their post-traumatic growth when they are facing a crisis (see the case studies of Nimali and Sarah, Chap. 9). For example, psychoeducation about the autonomic hierarchy helps normalize clients' challenges and responses to environmental cues (neuroception). Whether survivors struggle with PTSD or are rebuilding their lives in post-traumatic growth, it is necessary to remember that they have accumulated memories of trauma and unforgettable losses that will be triggered at times. Therefore, it is not surprising to see survivors who might otherwise develop post-traumatic growth occasionally struggle with life challenges (see Chaps. 8 and 9).

Acknowledging clients' emotional setbacks motivates them to seek support and rebuild their lives. While in practice, I met survivors who often asked, "I have been doing well, so why am I now falling apart because of this?" For these clients, psychoeducation restores the needed self-understanding. It creates calm (homeostasis) and, with the added process of co-regulation, rebuilds their confidence, refuels their inner strength and resilience, and allows them to re-ignite their post-traumatic growth.

In psychoeducation, attuned listening increases your opportunities to engage with your client, enabling them to benefit from co-regulation and enhance their ability to self-regulate and balance their emotions. Focused listening allows you to instinctively respond to what the client is saying and ask questions for clarification. Remember, individuals or groups displaying signs of post-traumatic growth possess inner strength, resilience, and a supportive network they can draw upon to overcome the temporary challenge that prompted them to seek services. It is also crucial to initiate the conversation about knowledge sharing, not as an expert on trauma intervention, but as a collaborative and supportive therapist open to discussion.

Providing psychoeducation helps you to

- Establish a positive relationship with your client that enables sensitive issues, including taboo subjects, to surface for discussion. This will help you build a connection while providing co-regulation.
- Clients experiencing post-traumatic growth seek help now because they cannot resolve issues as they did before; they are facing temporary setbacks that can be overcome (see the case studies in Chap. 9).

Inquire to determine where the client currently stands in the hierarchy of autonomic response. During a personal crisis, individuals undergoing post-traumatic growth (PTG) react to perceived threats by transitioning from social engagement (ventral vagal) to mobilization (sympathetic/fight-or-flight) and, when that is ineffective, to immobilization (dorsal vagal/shutdown). Pose questions about their recent crisis and available support system, utilizing a traumagram to re-ignite their PTG responses (Fig. 4.1).

- Evaluate the source of your client's post-traumatic growth and the effects of the recent crisis that pushed them back into their previous survival mode (see the case studies in Chap. 9—Nimali and Sarah). Provide information as necessary without judgment or adopting an expert demeanor.
- Pose questions out of curiosity. Ensure that the information you provide is culturally informed. Foster discussion, encourage your clients to make decisions, and engage them in solution-focused conversations.
- Explain the history of your client's post-traumatic growth responses and actions.

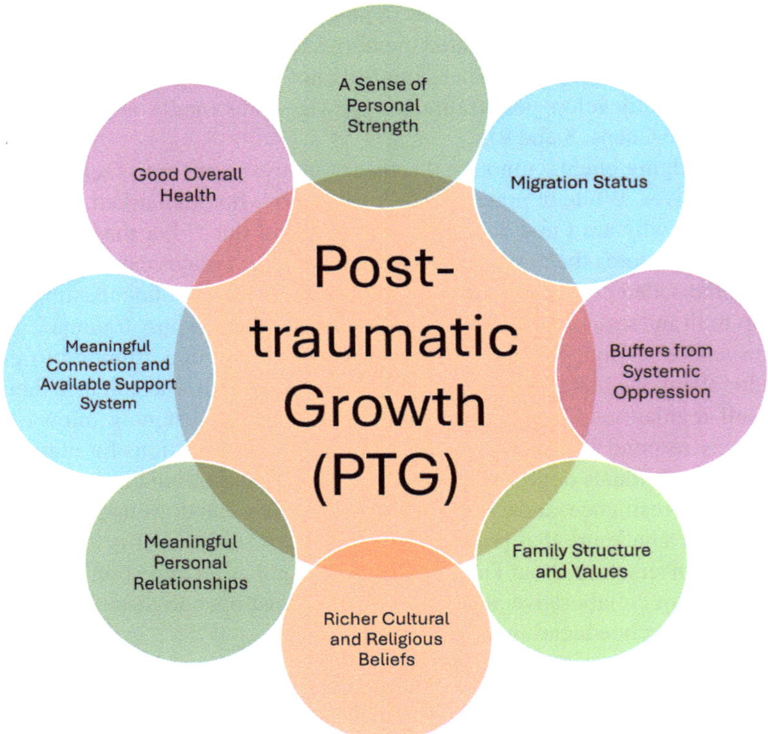

Fig. 4.1 Traumagram of support systems contributing to post-traumatic growth

- Identify past instances when the current issue did not pose difficulties for your client and discuss why this has changed. This process helps clients recognize adaptive stress responses, implement solutions, overcome their present challenges, and develop strategies for future ones.
- Collaborate to identify behaviors, emotions, and thoughts regarding the crisis and your client's meaning-making process.
- Collaboratively identify previous adaptive behaviors that fostered a positive outlook and the issues contributing to the current distress.
- Normalize the setback from PTG caused by the crisis to alleviate stress and self-doubt.
- Offer necessary support, including advocacy, to alleviate stress.
- Review your client's past coping skills to instill confidence and confirm that they still possess those abilities to help them navigate the current crisis.
- Listen to your client's survival story and rephrase it to increase their awareness of inner strength.

Helping clients develop self-monitoring and self-awareness skills often leads to positive change. Psychoeducation and self-monitoring activities enhance clients' awareness of their thoughts and actions. Maintaining a journal allows clients to document their thoughts, feelings, and behaviors, helping them recognize critical self-statements during crises that may affect their post-traumatic growth progress.

From personal experience, I understand that while I benefit from my post-traumatic growth during stressful times, I am also haunted by memories of life-threatening situations and severe injuries from the early stages of my migration. Occasionally, stressful life situations trigger memories that remain vivid even decades later, leading to sleepless nights and nightmares. Some days, I also face flashbacks, digestive issues, and increased sensitivity to danger. Remember, before their arrival, your clients may have experienced overwhelming trauma and might sometimes relive their nightmares as I do. If you listen, you will hear their devastating stories, triggered by current stressors and drawn from their accumulated memories of the past, including experiences of violence, sexual abuse, and physical harm. When this happens, I hope my story and those of others will help you, the student and social worker, understand the process of trauma, including the development of PTSD and PTG (see Chap. 3).

6 Critical Practice Considerations

In social work intervention, it is essential to identify the sources of your client's strengths and resiliency and the mechanisms that aided their development of post-traumatic growth. The following considerations will help you evaluate this in the context of planning an intervention (see Chap. 10 for further intervention practices).

Discussion Questions

1. What is your client's experience with trauma, and what is the origin of their post-traumatic growth?
2. What is the role of resilience in the development of post-traumatic growth? Using the traumagram in Fig. 4.1, explore the sources of resilience and growth in your client's life.
3. Define the reasons, timing, and methods for delivering psychoeducation.
4. When working with clients facing a recent crisis who have previously shown signs of post-traumatic growth (PTG), please pay attention to their current struggles or extreme stress, as this may be the source of their setback. What psycho-educational information can you provide to assist them during their current crisis?
5. Name at least two key concepts about post-traumatic growth that you learned from this chapter, and describe how you can apply them in your practice.

7 Activities to Consider

- Ask survivors who helps them manage their emotions and enhance their self-regulation, who contributes to their stress, and why.
- When emotions run high in a therapeutic session, social workers often utilize co-regulation by creating a safe and supportive environment to calm or stabilize survivors' nervous systems and emotions. Some skills they practice include attuned listening, respectful communication (mirroring/validation of emotions), and breathing or grounding techniques. Reflect on a moment when you applied a different skill and assess what worked and what did not in providing co-regulation.

Definitions

Co-regulation occurs when the nervous systems of two individuals interact, supporting each other and emotionally connecting through the mutual behavioral regulation of their physiological states. This phenomenon is linked to mirror neurons, which are the brain's natural mirror of the emotions and behaviors of others and are critical for social interaction and emotional understanding.

Medial orbitofrontal cortex (mOFC): This part of the brain is responsible for learning, memory, emotion, reward-related decision-making, and reinforcement. An association exists between post-traumatic growth (PTG) and increased activity in the left frontal cortex and the anterior regions of the mOFC, which may influence brain function and resilience.Competing InterestsThe author has no conflicts of interest to declare that are relevant to the content of this chapter.

References

Abraham, R., Lien, L., & Hanssen, I. (2018). Coping, resilience and posttraumatic growth among Eritrean female refugees living in Norwegian asylum reception centres: A qualitative study. *International Journal of Social Psychiatry, 64*((4), 359–366. https://doi.org/10.1177/0020764018765237

Ai, A. L., Tice, T. N., Whitsett, D. D., Ishisaka, T., & Chim, M. (2007). Posttraumatic symptoms and growth of Kosovar war refugees: The influence of hope and cognitive coping. *The Journal of Positive Psychology, 2*(1), 55–65. https://doi.org/10.1080/17439760601069341

Bem, D. J. (1967). Self-perception: An alternative interpretation of cognitive dissonance phenomena. *Psychological Review, 74*(3), 183–200. https://doi.org/10.1037/h0024835

Blechman, E. (2000). Resilience. In A. E. Kazdin (Ed.), *Encyclopedia of psychology* (Vol. 7). American Psychological Association.

Calhoun, L. G., & Tedeschi, R. G. (2014). Handbook of posttraumatic growth : research and practice (1st ed.). *Psychology Press*. https://doi.org/10.4324/9781315805597

Cozolino, L. J. (2010). *The neuroscience of psychotherapy: Healing the social brain* (2nd ed.). W. W. Norton & Company.

Fosha, D., Siegel, D. J., & Solomon, M. F. (2009). *The healing power of emotion: Affective neuroscience, development, & clinical practice* (1st ed.). W.W. Norton.

Fujisawa, T. X., Jung, M., Kojima, M., Saito, D. N., Kosaka, H., & Tomoda, A. (2015). Neural Basis of Psychological Growth following Adverse Experiences: A Resting-State Functional MRI Study. *PloS One, 10*(8), e0136427. https://doi.org/10.1371/journal.pone.0136427.

Hadad, M. (2013). *Satisfying life: Positive psychology and personal growth*. Pearson Canada.
Ledoux, J. (1996). *The emotional brain: The mysterious underpinnings of emotional life*. Simon & Schuster.
Levine, P. (2015). *Trauma and memory: Brain and body in a search for the living past: A practical guide for understanding and working with traumatic memory*. North Atlantic Books.
Mattson, E., James, L., & Engdahl, B. (2018). Personality factors and their impact on PTSD and post-traumatic growth is mediated by coping style among OIF/OEF veterans. *Military Medicine, 183*(9–10), e475–e480. https://doi.org/10.1093/milmed/usx201
Nakagawa, S., Sugiura, M., Sekiguchi, A., Kotozaki, Y., Miyauchi, C. M., Hanawa, S., Araki, T., Takeuchi, H., Sakuma, A., Taki, Y., & Kawashima, R. (2016). Effects of post-traumatic growth on the dorsolateral prefrontal cortex after a disaster. *Scientific Reports, 6*(1). https://doi.org/10.1038/srep34364
Nes, L. S., & Segerstrom, S. C. (2006). Dispositional optimism and coping: A meta-analytic review. *Personality and Social Psychology Review, 10*(3), 235–251. https://doi.org/10.1207/s15327957pspr1003_3
Ochsner, K. N., Bunge, S. A., Gross, J. J., & Gabrieli, J. D. E. (2002). Rethinking Feelings: An FMRI study of the cognitive regulation of emotion. *Journal of Cognitive Neuroscience, 14*(8), 1215–1229. https://doi.org/10.1162/089892902760807212.
Porges, S. W. (2007). The polyvagal perspective. *Biological Psychology, 74*(2), 116–143. https://doi.org/10.1016/j.biopsycho.2006.06.009
Porges, S. W. (2017). *The pocket guide to the polyvagal theory: The transformative power of feeling safe*. W. W. Norton.
Porges, S. W., & Dana, D. (2018). *Clinical applications of the polyvagal theory: The emergence of polyvagal-informed therapies* (Norton series on interpersonal neurobiology). W. W. Norton & Company.
Prati, G., & Pietrantoni, L. (2009). Optimism, social support, and coping strategies as factors contributing to posttraumatic growth: A meta-analysis. *Journal of Loss & Trauma, 14*(5), 364–388. https://doi.org/10.1080/15325020902724271
Saleebey, D. (2013). *The strengths perspective in social work practice* (6th ed.). Pearson.
Schore, A. N. (2002). Advances in Neuropsychoanalysis, Attachment Theory, and Trauma Research: Implications for Self Psychology. *Psychoanalytic Inquiry, 22*(3), 433–484. https://doi.org/10.1080/07351692209348996
Shaw, A., Joseph, S., & Linley, P. A. (2005). Religion, spirituality, and posttraumatic growth: A systematic review. *Mental Health, Religion, & Culture, 8*, 1–11. https://doi.org/10.1080/1367467032000157981
Siegel, D. J. (2007). *The mindful brain: Reflection and attunement in the cultivation of Well-Being*. Norton & Company.
Siegel, D. J. (2010). *The mindful therapist: A clinician's guide to mindsight and neural integration* (Norton series on interpersonal neurobiology). W. W. Norton & Company.
Siegel, D. J. (2011). *Mindsight: The new science of personal transformation*. Bantam.
Simsir, Z., & Dilmac, B. (2018). Predictive relationship between war posttraumatic growth, values, and perceived social support. *Illness, Crisis, and Loss/Illness, Crisis & Loss, 29*(2), 95–111. https://doi.org/10.1177/1054137318788655
Tedeschi, R., & Calhoun, L. G. (1995). *Trauma and transformation: Growing in the aftermath of suffering*. Sage.
Tedeschi, R. G., & Calhoun, L. G. (1996). The posttraumatic growth inventory: Measuring the positive legacy of trauma. *Journal of Traumatic Stress, 9*(3), 455–471. https://doi.org/10.1002/jts.2490090305
Tedeschi, R. G., & Calhoun, L. G. (2004). TARGET ARTICLE: "Posttraumatic growth: Conceptual foundations and empirical evidence". *Psychological Inquiry, 15*(1), 1–18. https://doi.org/10.1207/s15327965pli1501_01
Tedeschi, R. G., Shakespeare-Finch, J., Calhoun, L. G., & Taku, K. (2018). *Posttraumatic growth: Theory, research, and applications*. Routledge.

Uy, K. K., & Okubo, Y. (2018). Reassembling a shattered life: A study of posttraumatic growth in displaced Cambodian community leaders. *Asian American Journal of Psychology, 9*(1), 47–61. https://doi.org/10.1037/aap0000111

Van Der Kolk, B. A. (2014). *The body keeps the score: Brain, mind, and body in the healing of trauma*. Penguin Publishing Group.

Part III
The Challenges of Trauma Care

Chapter 5
Understanding the Causes of Service Gaps

1 Social Work Practice and Trauma Care

Social workers practice in multiple settings, providing services such as resettlement and mental health interventions. Those new to the field find themselves overwhelmed by the need for practice skills to provide trauma-informed, culturally based intervention (TICBI) as they become the go-to person in their agency, providing trauma care. As practitioners guided by anti-oppressive practice (AOP), freshly minted social workers carry stress from seeing repetitive trauma patterns in survivors because of service gaps.

Despite these challenges, AOP in social work remains a helpful guide, particularly for those with egalitarian beliefs and social justice values, in avoiding service gaps as much as possible. As dedicated advocates for social care, social workers using a client-centered approach seek solutions to the issues clients face, which is necessary to prevent service gaps for survivors and their own burnout. In recent years, conversations among practitioners have shown the impact of increasing job demands on social workers and how their workload affects their service provision. The social workers who participated in my TICBI study often found the work of providing trauma care personally rewarding. The job demands for those involved in trauma intervention are sometimes overwhelming, affecting their well-being. Therefore, I asked service providers about their day-to-day challenges in trauma care and the source of their internal strength and resiliency that keeps them continuing the work.

N. Bokore, *Trauma-Informed, Culturally Based Intervention*,
https://doi.org/10.1007/978-3-032-06835-4_5

2 Service Providers' Stories

2.1 Intervention Methods

I began my TICBI study looking for answers to practice-based challenges and the blanket use of the Western one-size-fits-all intervention model taught in graduate schools when working with multicultural, multiethnic, and multi-issue clients. The study explores these service challenges, and we developed a collaborative intervention model for inclusive practice. I received funding from the Social Sciences and Humanities Research Council (SSHRC) before the COVID-19 pandemic in 2020 and began the TICBI study online during the pandemic. It was stressful as we transitioned to working, teaching, and doing research online. Despite these challenges, I recruited 20 service providers who participated in two focus groups (two groups of eight participants) and four individual interviews. Since the service providers provided most of their service online during the pandemic and dealt with social distancing, I began the focus group discussions with a general conversation about their understanding of trauma and trauma care in an AOP practice committed to cultural awareness, equity, and social justice. I asked questions about service gaps and using TICBI in their everyday practice.

Despite the variation in training backgrounds, focus group participants provided surprising insights, such as "You can see how the distressing event impacts clients." For some, the experience creates feelings of hopelessness and mistrust, which affect their resettlement journeys. When asked about trauma care and service gaps, discussions in each focus group began with an overview of their current service models. One participant remarked, "Trauma care involves listening to survivors' stories and increasing opportunities for groups to benefit from social gatherings and share their experiences, as well as providing resources to reduce stigma and isolation." Most participants described the benefits of social gatherings in promoting their clients' healing processes. They also discussed how they organize these gatherings to foster conversations based on survivors' shared experiences (such as political situations and countries of origin) and their trauma responses. They identified group gatherings and counseling as valuable for clients and a means to enhance their advocacy skills to effect systemic change and improve their clients' living conditions.

When the service providers were asked about the methods or theories informing their practice, they responded differently based on their educational backgrounds. Those with graduate degrees practicing psychotherapy discussed how they integrate culture into their trauma-informed practice. Those working in resettlement services emphasized that since they work with clients with diverse needs, AOP helps them overcome challenges in trauma care and service gaps. Both groups explained how they use AOP to analyze the overlapping systemic structures that make survivors vulnerable to marginalization and service gaps. This increased their motivation to continue to work and help those needing their services. Some participants explained how they use AOP concepts in their person-centered practice to challenge inequalities, including the service gaps encountered by vulnerable groups they serve. They talked about how the explanatory function of AOP helps them provide trauma care in the context of survivors' lived experiences, helps them acknowledge the diversity of their clients' service needs, and encourages them to advocate for change in the face of funding-related challenges. Quoting Dominelli (2002, p. 13), one focus group participant said AOP and culturally based trauma care is "the creation of non-oppressive systems of care."

In these discussions, focus group participants conferred on the impact of systemic structures that create barriers for service providers and clients. One service provider explained how, to overcome these barriers, she collaborates with survivors as the starting point; she focuses on creating a connection with them before beginning the actual intervention: "I use what I call a warm transfer process where I focus on rapport building, engaging clients in culturally acceptable communication." Another service provider described the impact of trauma during resettlement and the challenges of navigating structural issues, saying how it keeps survivors in an "emotional and physical shock that keeps them stuck at the lower level of their vagal response as they continue to be re-traumatized by the struggle with existing barriers." She also described the theories and models informing her intervention process to help clients regulate their emotions as a combination of "cognitive-behavioural therapy, mindfulness, and acceptance and commitment therapy." Acceptance, for this service provider, "is not a denial of the past or giving up and not doing anything (freeze), but allowing oneself to accept the wound as part of one's personal experiences," with which other members agreed.

As they learned about each other's practice models, participants in each focus group discussed the importance of considering the client's culture, the variation in trauma responses, and clients' willingness to address subjects typically reserved for family discussion (Chaps. 8 and 9). The focus group participants also connected cultural communication with regulating survivors' emotions (hyperarousal or hypoarousal) and widening their "window of tolerance," the optimal zone or arousal state for normal function to manage daily stress and enhance coping mechanisms (Siegel, 2011). They also examined how the client's culture shapes the pace of the conversation and their assessment of the client's reactions to events and their impacts on their lives. In response to questions about integrating TICBI and AOP in their services to bridge service gaps, they emphasized the significance of collaborative work, including psychoeducation, sharing information, and providing advocacy.

In the individual interviews, service providers raised similar discussion points about the skills they utilize to provide trauma care. One participant emphasized the importance of being "attuned to clients' emotions as they share difficult subjects," demonstrating support and empathy for what they are experiencing and applying specific theories to explain the issue to the client or in planning change. One participant discussed how they employ Steven Hayes's (2015) acceptance and commitment therapy (ACT) to introduce mindfulness (see Chaps. 9 and 10), using appropriate language, cultural metaphors, and values. Another participant highlighted the significance of culturally based communication in trauma care, the client's core values, and concentrating on clients' actions and behaviors. Providing

TICBI involves refocusing clients' traumatic stories toward actions that promote growth (see Chap. 4).

Similarly, individual participants working in resettlement services described how they based their work in AOP and managed the overwhelming caseloads created by funding shortages and the ever-decreasing number of service providers; they prioritized the process of helping and their core values. They shared their dedication to the field, discussing how they handle their daily work using individual self-care methods: seeking social support and staying connected to family and friends. They emphasized the importance of having support systems, especially after hearing clients' shocking stories, coping with their lack of training for specific service needs, and the shortage of specialized care to refer their clients.

Addressing the helpful skills they use, one service provider discussed how embracing "difficult conversations at the client's pace helps them and their clients." This process decreased their stress and helped them focus on solving the client's pressing issue. Skills such as "normalizing the situation the clients are experiencing and demonstrating empathy" were also expressed by this participant as part of their practice in trauma care. Participants in both the focus group and individual interviews emphasized the benefits of an AOP approach in trauma care, focusing on oppressive systems that create service gaps. Some noted the benefit of sharing a language and culture with clients, which made it possible to help them explain therapeutic methods better.

Participants in the study pointed out the careful steps needed to provide trauma care to avoid re-traumatization. One participant described their skill: "working at [a] client's pace and not overwhelming them by wearing the expert hat and having a one-sided conversation without considering the client's interpretation of the situation due to culture and capacity." Another participant said, "I used AOP-based trauma care, working with clients at their pace, using the language they use to describe what is going on in their lives to decrease the overwhelming stress related to sharing personal stories with strangers." These participants highlighted the challenges of providing services to those with similar backgrounds, as one explained, "Some survivors might avoid people [service providers] from their own culture because of past traumas and what they represent." Participants also talked about service gaps and service provider burnout brought on by funding cuts, language barriers, lack of interpreters, and heavy caseloads. They described the gap as a danger that can create miscommunication, misinterpretation, and misdiagnosis.

2.2 Service Gaps

Below are excerpts from the focus groups and individual interviews. When I started working over a decade ago, the caseloads were busy. However, the management team at the agency was somewhat understanding about our workload and the demands of the job and supportive. The current limitation on funding and the demand for monthly stats/reports on provided services pose challenges for the management team and us. Unfortunately, what I see now is making giving clients the emotional support and services they need increasingly difficult. That is where a lot of the burnout originates, because I am supposed to help. However, there are no accessible services, especially in mental health, housing, rising costs of living, and limited income.

Some service providers discussed how they contended with increasing caseloads, inadequate time allotment created by funding constraints, and the challenge of keeping up with the ongoing program and policy changes in the sector. Here is what one resettlement worker said while addressing service gaps:

> I have a large caseload, and my clients struggle with finding housing, employment, and addressing mental health issues. Finding affordable housing for new arrivals is now formidable, and employment is another challenge as they struggle with the initial resettlement and language learning. The need for mental health services is high, and as new arrivals are more open to discussing the issues, I struggle with finding services for them. So, my frustration and daily stress are about how I can help my clients because I am not a clinician and cannot help them find an affordable one.

The social workers in resettlement services discussed how they faced a moral dilemma that they deal with, which is between wanting to help, inadequate resources, and challenges when advocating for clients. One participant commented on the moral distress facing service providers:

> We are pressured to reach our service targets while dealing with large caseloads and higher needs. While clients need to be heard or at least [we need to be] a sounding board for their issues, we are pressured with a tight schedule of seeing clients only for 30 minutes or a maximum of less than an hour. This is highly stressful, mainly when working with traumatized clients, because I have to rush the session and then stop. After all, another client is waiting for me.

Social workers have an ethical responsibility to support one another, especially when they observe a colleague struggling with physical, mental, and emotional health issues that affect their work. Providing professional support in the workplace is crucial and challenging for helping colleagues who seem to be experiencing burnout. The challenge lies in the disconnect between professional values and personal struggles, with systemic structures contributing to service gaps. Nevertheless, it is essential to address these issues collaboratively and assist colleagues in need by highlighting the significance of self-care and well-being, thereby helping the colleague while simultaneously reducing the service gaps that might persist otherwise (Kleespies et al., 2011).

3 The Impact of Work-Related Stress

The probability of work-related stress is high in these challenging times, given the significant need for services and existing service gaps, especially when working with vulnerable populations. The working conditions and clients' traumatic stories increase the chances of vicarious trauma, which affects both physical and mental health (Selye, 1976; Szabo et al., 2012). As mentioned earlier, work-related stress and vulnerability to burnout are becoming everyday realities. The phenomenon of work-related stress first appeared in the literature in the early 1970s, initially described as compassion fatigue (Devilly et al., 2009), now also known as secondary traumatic stress and vicarious trauma, whose symptoms closely resemble those of PTSD and stem from accumulated work-related stress (McCann & Pearlman, 1990).

Some of the causes of secondary traumatic stress and vicarious trauma are also linked to bystanders' guilt, which creates feelings of shame and self-doubt and increases service providers' chances of burnout. Participants in my study, both in focus groups and individually, discussed how workload stress, additional responsibilities, and entering survivors' worlds through their stories connect to secondary traumatic stress and vicarious trauma. As they debated these types of stress and health risks, they remained dedicated to their profession due to their egalitarian and social justice values, which often encouraged them to continue their work.

Multiple studies have shown that since Hans Selye (1976) introduced work-related stress in 1936, its impact on health has been a significant topic (Levine, 2015; Newell & MacNeil, 2010; Ogden & Fisher, 2015; Oshri et al., 2022; van der Kolk, 2014). Defining the benefits and challenges of different levels (mild, moderate, or intermittent), researchers Oshri et al. (2022, p. 2) explained that these types of stress have no health impact unless the person has underlying neurobiological sensitivities that intersect with current work-related stress. Work-related stress is also subjective, based on a person's perception of what constitutes a threat and whether they hold an optimistic view supported by neurobiological and behavioral adaptive responses to acute and situational anxiety. However, these researchers caution about the impact of work-related stress that lies beyond a person's control (e.g., workload and lack of support), which can create a toxic response and increase the likelihood of burnout. While some service providers can tolerate mild, moderate, and intermittent stress in everyday work environments, failing to address it poses health risks for others (Selye, 1976). Therefore, paying attention to service gaps and the impact of work-related stress on the body and brain is essential. Self-awareness of your tolerance level, informed by various supportive factors including internal resilience, is critical; when this level becomes intolerable, seeking help is necessary.

3.1 Body/Brain Responses to Work-Related Stress

The impact of stress on the body/brain is related to the release of large amounts of glucocorticoid/cortisol (stress hormones) that continues long after the danger of the event has passed (Selye, 1976). The human adaptive response tolerates mild or moderate stress without leading to excessive fatigue (hormetic stress). However, sympathetic reactions occur when stress exceeds the body's capacity to cope (Oshri et al., 2022). For social workers, persistent stress and the excessive production of stress hormones that accompany it make them vulnerable to changes in their brain's neurochemistry (Oshri et al., 2022; van der Kolk et al., 1996). These changes influence the social worker's metabolism, affecting the breakdown of protein and its conversion to glucose, increasing the availability of fats for fuel or energy to enhance blood flow and stimulate survival responses, providing the energy needed to manage their increasing workload.

Overwhelming work stress has been shown to affect information integration in the central nervous system (CNS), which regulates neural activities in the body's stress response (van der Kolk et al., 1996). This process impacts the hypothalamic-pituitary-adrenal (HPA) axis, increasing the stress hormone cortisol or glucocorticoid production. Under normal circumstances, this response is necessary to manage day-to-day stress; however, it becomes problematic when persistent. Chronic production of stress hormones alters cognitive function and the ability to interpret stressful situations while also heightening the sympathetic (alert, fight, flight, or freeze) response (van der Kolk et al., 1996). Consequently, work stress adversely affects service providers' health by increasing the activation of their neuroception and danger alerts, or autonomic responses, impacting their overall well-being.

3.2 Unintentional Service Gaps

I have seen the impact of chronic stress brought on by excessive service demands and workloads. Stress affects the functions of the endocrine system, as mentioned above, including the release of hormones, and also impacts higher brain areas, affecting executive function (van der Kolk, 2014). This disrupts social workers' ability to manage daily stress, anticipate challenges, and know when to seek help, increasing their chances of burnout. The persistence of this type of stress also triggers primitive and self-protective behavior, affecting the collaborative work required in trauma care and adding to stress, self-blame, and burnout (Cozolino, 2010; Selye, 1976; Szabo et al., 2012; van der Kolk, 2014).

3.3 Preventing Unintentional Service Gaps

Prevention begins with advocacy to reduce workplace stress and create healthy work environments for social workers. Social workers who tirelessly strive to provide accessible support systems for their clients understand that having social support for themselves is crucial to stimulating their cell regrowth and neuroplasticity. This enhances their resilience and reconnects brain areas affected by service-related stress and the pressures of structural changes. Increasing self-care and recognizing the impact of social isolation in the workplace hinders the body–brain processes, leading to internal damage that ultimately affects well-being.

It is a point overwhelmingly agreed upon by the participants in my study, highlighting the importance of self-care even when it is not widely accessible, such as utilizing available support systems, like colleagues, and engaging in culturally based mindfulness exercises (see Chap. 10). One of the suggested self-care tips is to walk a short distance after feeling stressed during breaks or after work to relieve muscle and emotional tension. Participants also noted that job demands often make these activities impossible for some service providers. When I asked for other tips, they mentioned engaging in activities during their off-work time, such as joining cultural dance groups, yoga, and tai chi, to reduce stress and minimize hormonal production.

Drawing on the ever-growing knowledge of social neuroscience, social workers can access information about body and brain reactions to stress responses to prevent burnout, avoid other work-related health risks such as vicarious trauma, and support colleagues experiencing it.

4 Trauma Care and the Challenge of Stress

Social workers assist the most vulnerable individuals, families, and groups affected by trauma. Amid the current state of social programs and funding cuts, there is an increasing demand for social workers to deliver care despite limited time and resources. Nevertheless, this means that those providing trauma care may confront the complex challenges of secondary traumatic stress, vicarious trauma, compassion fatigue, and burnout (Kase et al., 2018; Whitt-Woosley & Sprang, 2023).

More than ever, social workers involved in community programs face increasing workloads, including serving numerous clients and maintaining meticulous recordkeeping daily. The demands for services from communities are rising for agencies, particularly those offering resettlement services, which require shift and weekend work and availability for emergency calls. This leaves little time for self-care through social interaction with colleagues.

Social workers are social beings who share their stress, internal feelings, and pressure from external events such as policy and funding changes. Sharing emotions and energy with colleagues allows them to benefit from co-regulation. Social workers constantly synchronize with others' energy and feelings; they share emotions and feelings through their mirror neurons as they assess and evaluate their clients' needs or work to understand their feelings and actions. One of the greatest gifts of the social brain is that self-regulated social workers can help regulate clients' emotional dysregulation through natural and straightforward neurological echoes of mirror neurons and behaviors based on positive and nurturing social exchange (Cozolino, 2010; Siegel, 2011).

Some social workers are affected more adversely than others by their early life experiences. This diminishes their stress responses or distorts their views. Self-reflection and seeking support are needed to prevent professional criticism and vulnerability to compassion fatigue, secondary traumatic stress, and vicarious trauma. Dealing with heavy caseloads with a large number of clients needing TICBI increases the likelihood of vicarious trauma. Vicarious trauma first impacted psychotherapists in the early 1990s (Whitt-Woosley & Sprang, 2023). Since then, other service providers such as police officers, 911 operators, social workers, and nurses have also experienced it.

Social workers, as part of the service provider team, experience work-related vicarious trauma, also known as secondary trauma. They encounter the emotional and psychological distress of their clients, especially when working with individuals affected by prolonged trauma. This exposure can result in various adverse consequences, including burnout, which impacts their well-being. Other work-related symptoms for social workers include self-isolation and emotional withdrawal from colleagues, friends, or activities they previously enjoyed. This upsets their emotional balance and removes them from the solace they need to continue providing trauma care. Increased feelings of sadness, anxiety, fear, and vulnerability to making mistakes arise from their struggle to maintain focus, which heightens their anxiety, irritability, and fear, and leads to greater isolation. Ultimately, these symptoms add to their stress, affecting their well-being (Cozolino, 2010).

5 Stress-Creating Factors in Social Work: Transference and Countertransference

Managing a heavy caseload of trauma survivors while being overworked contributes to compassion fatigue, which is linked to work-related stress and insufficient support. This stress leads to emotional and physical exhaustion, a decreased capacity for empathy, and feelings of burnout. Additionally, secondary trauma and burnout heighten social workers' vulnerability to setting boundaries between their clients' emotions and their own, which can lead to countertransference or the need to manage a transference-related case as it arises. Therefore, being aware of other work-related challenges, such as transference and countertransference, is crucial for delivering trauma-informed services.

In the late nineteenth century, Freud first discussed transference and countertransference, describing how clinicians and clients experience each other's emotions within their therapeutic relationship. Psychodynamic theorists explain countertransference as the clinician's unconscious and negative perceptions of the client (Ellis et al., 2018). These perceptions stem from the clinician's negative past experiences (social exchange theory), impacting their clinical interaction and empathic response, which may contribute to the symptoms of compassion fatigue.

Transference represents another practice-based challenge that arises from unconscious relational interactions. It is associated with early life experiences (refer to Social Interaction Theory and the work of George Herbert Mead from the 1920s to the 1930s) and unresolved relational trauma. Overburdened social workers addressing these early life experiences may find themselves re-experiencing their own early life and adaptive survival experiences. Such re-experiencing can manifest as positive or negative transference, influencing session interaction and the therapeutic process. In either case, positive and negative transference significantly impact clinical intervention and trauma care. For instance, positive transference may facilitate interaction when clients connect with their clinicians, regard their knowledge highly, value the care they receive, and see the support as beneficial to their recovery.

On the other hand, it may create transference and dependency, which can impact the planned change. Positive transference occurs when clinicians become overprotective of clients and work too hard to resolve their pressing issues. These clients often remind clinicians of their past or of someone to whom they had a strong attachment. Similarly, negative transference influences interactions as clients unconsciously redirect their feelings of unhappiness, anger, or abuse toward the therapist, altering the communication dynamic because clients perceive clinical actions as dangerous or threatening, which can increase their fight-or-flight response.

Another challenge of transference and countertransference that arises when providing TICBI is the cultural misinterpretation of the interaction. This misinterpretation affects the evaluation of emotional bonds that both the client and the service provider need to foster change talk. To prevent cultural misinterpretation from impacting emotional interactions and creating transference, "the use of self" is essential in social work practice. It enables social work skills and personal characteristics to make the interaction authentic and value-based. Social workers frequently engage in self-reflection to assess how they navigate their professional knowledge, values, and skills while providing care.

The use of self in trauma care is crucial for preventing vicarious trauma or burnout, as social workers depend on self-knowledge to attune to their internal and implicit working models, including evaluating their neuroception of danger. It is essential to acknowledge countertransference, as it amplifies danger-detecting neuroception and triggers adaptive and survival-based responses, thereby affecting social workers' professional relationships with their clients. To alleviate this, stressed service providers should regularly self-reflect, engage with cultural brokers, and surround themselves with supportive colleagues and friends to prevent secondary traumatic stress or vicarious trauma while also fostering vicarious resilience and vicarious post-traumatic growth.

6 Developing Vicarious Resilience and Vicarious Post-Traumatic Growth

While some service providers are affected by vicarious trauma, others, as they provide trauma care, develop what is known as vicarious resilience, or personal growth and a positive outlook on their professional life. Vicarious post-traumatic growth is not unlike post-traumatic growth (see Chap. 4). The development of vicarious post-traumatic growth is linked to service providers' internal capacity for self-awareness and self-care and their resourcefulness in harnessing support that helps them overcome work-related stress and challenges (Puvimanasinghe et al., 2015). For some social workers, witnessing clients' resilience allows them to find the strength to change their own lives, altering their perceptions of interpersonal relationships and their philosophy of life (Kalaitzaki et al., 2024; Puvimanasinghe et al., 2015; Tedeschi & Calhoun, 1995; Tedeschi et al., 2018).

Listening to others' trauma stories is a challenging task that requires self-care and support. Self-awareness and the ability to pay attention to one's internal feelings enhance the likelihood of developing vicarious post-traumatic growth (VPTG), listening with compassion, and strengthening the working relationships necessary for change talk in TICBI. VPTG refers to the positive changes and growth that service providers experience as they listen to the trauma narratives of their clients, leading to indirect exposure to trauma. Social workers and other professionals occasionally undergo VPTG, which results in increased empathy, a stronger professional identity, and a deeper understanding of the human experience (Brockhouse et al., 2011; Cohen & Collens, 2013). For some social workers, this experience gives them a new perspective and renewed strength to serve those impacted by trauma (Engstrom et al., 2008). This, in turn, facilitates vicarious growth as their trauma work imbues meaning into their work, redefines their worldview, and positively alters their work identity (McCormack et al., 2011).

The social workers who were part of my TICBI project discussed vicarious post-traumatic growth and what they learned from their clients as they interacted and exchanged knowledge and wisdom, mutually affecting each other's views and perspectives. Similarly, when I worked with traumatized clients, I struggled with a flashback of exhaustion and sometimes dreaded taking on another client. One day, I had lunch with a colleague; while talking about work, she shared with me her past traumatic experiences and how she survived in her personal life and managed her professional obligation to work with traumatized clients with similar experiences. Hearing her story gave me great admiration for her resiliency. In that short lunch meeting, she shared helpful information about self-care and self-regulation and gave me the emotional support I needed. She became my go-to colleague when I needed support. She became a reliable support system when I needed one, contributing to the development of my resilience and growth. A friendly lunch meeting and an exchange of skills wrapped up in an advice package made a difference in my clinical career, changing my belief about my abilities and knowledge and what I could contribute to others' lives as a professional. Social workers are bound by their commitment to helping others in need. However, they also provide each other with a social support system, increasing their neuroception of safety, promoting engagement, connection, and self-care, and contributing to their vicarious post-traumatic growth.

6.1 Self-Care Strategies

Caring for oneself is as important as caring for others. In providing services to the vulnerable, social workers are vulnerable to vicarious trauma and other work-related and stress-induced health risks. Organizational support provides overworked social workers with simple peer-based support systems that can be operative during debriefing sessions. Check on those dealing with heavy caseloads and offer them self-care opportunities such as lunch, yoga, or meditation. These activities can reverse the activation of their autonomic nervous functions from fight, flight, or freeze to relaxation, centeredness, the activation of their ventral vagal system, and access to their cognitive processing system (see Chap. 9). The social worker's autonomic nervous system functions hierarchically in stressful work conditions, monitoring the environment and regulating body/brain responses according to workplace stress levels. For example, when social workers receive support or participate in activities that decrease their stress and activate their ventral vagal system, they can focus on creating a work-life balance, which helps them cope with the stress of work. In contrast, when social workers are dealing with heavy workloads and are overwhelmed by multiple stressors, they may feel anxious, angry, or irritated. Their survival response is based on the sympathetic nervous system telling them about the impact of their work on their body and brain.

Further signs of a depilated survival system are evident when social workers are no longer able to tolerate the burden of their workload or have difficulty expressing their feelings in words (verbal amnesia). Working in an unsupportive work environment increases fight-or-flight responses and creates a sense of being stuck in never-ending stress with no way out. When this continues, they show signs of chronic burnout and emotional disturbances (cynical, detached, or self-doubt), physical symptoms (frequent physical illness and absenteeism from work), and behavioral change (inadequate or low work performance, procrastination, or anger).

Attention to mental and physical exhaustion is essential in assessing social workers' work-life balance. Working with traumatized clients requires prioritizing the social worker's well-being. Minimizing their stress levels and providing them with a stable and sustainable workload are musts in delivering effective TICBI. Social workers must use their advocacy skills to achieve work-life balance and continue helping survivors.

- First, seek institutional or agency support to create a work-life balance. This support system will restore your spirit, re-energize your strength and commitment, re-ignite your compassionate self, activate your ventral vagal system, help you reconnect with your clients, and provide efficient services.
- Second, explore individualized self-care strategies that work for you, including stress-reduction activities such as socialization, exercise, and meditation.
- Below are my self-care strategies. However, a social worker can look for other methods specific to their interests, personalities, and culture. For social workers, identifying personal strategies can help restore energy and confidence, increasing their ability to tolerate work-related stress. In trauma care, paying attention

to stress-related sensations and taking action are preventative steps to avoid burnout, compassion fatigue, and service gaps. Therefore, when providing trauma care,

- Tune in to yourself, paying attention to your sense of safety and internal emotions.
- Note the presence of transference or countertransference during client interactions. Track your client's emotions in each interaction using daily personal case notes as a source of self-reflection.
- In addition to completing case notes after each session, reflect on the theories or methods you used for your own personal growth.
- Create appropriate boundaries and self-care, as you work on rapport-building for collaborative work.
- Pay attention to your rescuing behavior, or the "professional side effect." It is important to remember that as a social worker, you are also a human with a history and personal struggles. Some social workers overdo the work of helping, showing excessive compassion and avoiding confrontation, which exacerbates their stress and thus inadvertently impacts their health. Pay attention to your self-sacrificing behavior and check in with colleagues and supervisors for feedback on making a healthy change.
- Make time to nurture your spirit by participating in enjoyable activities, but avoid mind-altering activities (alcohol, drugs, and other unhealthy behaviors). Pursue activities that allow you to exchange exhaustion for physical and mental nourishment.
- Participate in activities that promote mindfulness based on spirituality and culture. Living in a chaotic work environment sometimes leads to a loss of identity, and these activities help reactivate the autonomic response and the ventral vagal nervous system. Religious and cultural activities strengthened my identity.
- Connect with like-minded people who will support you and help you promote structural and social change so that you can offer the needed programs and services.
- Look for opportunities to learn new and healthy after-hours activities. This process will create new neural connections, improve energy, and give you a sense of accomplishment.
- Seek opportunities for supervision. Clinical supervision will allow you to increase your strengths, develop new skills, and grow professionally. Clinical individual or group supervision provides a good client–therapist relationship, making a difference in your well-being. If there are no opportunities for supervision, advocate for it in your agency.

6.2 Supporting Clients and Social Workers through Advocacy

One of social work's disciplinary strengths is its understanding of how advocacy creates change. Taking note of Maurice Moreau's (1979) interdisciplinary work in defining the structural approach, social workers can work toward transforming current oppressive institutional structures that normalize heavy workloads to decrease the financial costs of community programs or the demand for more work for less pay. Healthy workplaces and social relations should govern the services we provide to prevent service gaps.

Advocating for organizational involvement is also needed to ensure resources are available during challenging times. Organizations that provide services for newcomers, particularly survivors of war and forced migration, should increase their support for their staff and clinicians who are providing trauma-informed interventions. Supervision, peer supervision, or mentorship should be made available for everyone (from frontline workers to clinicians doing psychotherapy). The availability of these programs for social workers will contribute to their coping and resilience-building mechanisms required for social work practice. Resilience in social work is broadly understood as doing better than expected in the face of adversity. It should be used in the workplace to promote health, where those working with traumatized individuals can access resilience-building resources such as educational training and organized social events to nurture collegial support, create connections, and strengthen relationships. Other helpful practices:

- Check how those working with survivors balance personal and private lives to encourage more robust support systems.
- Create opportunities for peer support and consultation within the agency.
- Openly discuss learning needs, self-assessment, and work-life priorities during meetings.
- Provide ongoing professional development seminars, workshops, and skill training to improve services.

Definitions

Compassion fatigue and burnout: Often confused in social work. Burnout is a psychological condition caused by the chronic relational stress of being overworked without support. Compassion fatigue is a consequence of working with traumatized individuals and impacts social workers' health when they are overworked or overwhelmed by the stories of others, increasing their vulnerability. Compassion fatigue is different than secondary traumatic stress and vicarious trauma. The symptoms of compassion fatigue include

- Feeling helpless or powerless during a session, and sensing fear or self-doubt during sessions.
- An inability to resonate with and show empathy and compassion toward clients as they tell their stories causes a disconnection with your client.
- Feeling overwhelmed, exhausted, and unable to fulfill work demands.
- Feeling emotional detachment, numbness, intense emotional reactions, or difficulty providing the emotional connection needed during a session.

Vicarious or secondary traumatic stress, as some call it, refers to the indirect exposure to the trauma of others or the emotional pressure experienced when hearing individuals or families recount their traumatic experiences. For social workers, this is an occupational hazard, as they routinely encounter survivors sharing their stories of traumatic events. Repeatedly hearing their clients' accounts of trauma increases their vulnerability to compassion fatigue and burnout. Symptoms of secondary traumatic stress are similar to those of PTSD.

Vicarious post-traumatic growth is related to the development of post-traumatic growth and post-traumatic resilience. It happens when therapists are positively affected by their clients' resiliency in the face of trauma. Hearing clients' stories and witnessing their actions encourages social workers to appreciate human resilience and increase their confidence that their work is helpful despite systemic challenges. This helps social workers maintain their vicarious post-traumatic growth and avoid burnout. It motivates them to continue providing services, create needed programs, and increase their advocacy work to help survivors.

Discussion Questions

1. Social workers often discuss what some call vicarious or secondary traumatic stress in lunchrooms and staff meetings without elevating the issue to an organizational level. Explain why these discussions require deeper attention and advocacy efforts in the current work environment.
2. Discuss the advantages and disadvantages of addressing vicarious or secondary trauma among colleagues in your professional career or practicum placement positions.
3. Discuss prevention methods you can implement to promote your well-being and that of your colleagues while enhancing client care.Competing InterestsThe author has no conflicts of interest to declare that are relevant to the content of this chapter.

References

Brockhouse, R., Msetfi, R. M., Cohen, K., & Joseph, S. (2011). Vicarious exposure to trauma and growth in therapists: The moderating effects of sense of coherence, organizational support, and empathy. *Journal of Traumatic Stress, 24*(6), 735–742. https://doi.org/10.1002/jts.20704

Cohen, K., & Collens, P. (2013). The impact of trauma work on trauma workers: A metasynthesis on vicarious trauma and vicarious posttraumatic growth. *Psychological Trauma, 5*(6), 570–580. https://doi.org/10.1037/a0030388

Cozolino, L. J. (2010). *The neuroscience of psychotherapy: Healing the social brain* (2nd ed.). W. W. Norton & Company.

Devilly, G. J., Wright, R., & Varker, T. (2009). Vicarious trauma, secondary traumatic stress or simply burnout? Effect of trauma therapy on mental health professionals. *Australian and New Zealand Journal of Psychiatry, 43*(4), 373–385. https://doi.org/10.1080/00048670902721079

Dominelli, L. (2002). *Anti-oppressive social work theory and practice*. Palgrave. https://doi.org/10.1007/978-1-4039-1400-2

Ellis, T. E., Schwartz, J. a. J., & Rufino, K. A. (2018). Negative Reactions of Therapists Working with Suicidal Patients: a CBT/Mindfulness Perspective on "Countertransference." *International Journal of Cognitive Therapy, 11*(1), 80–99. https://doi.org/10.1007/s41811-018-0005-1.

Engstrom, D., Hernandez, P., & Gangsei, D. (2008). Vicarious resilience: A qualitative investigation into its description. *Traumatology, 14*(3), 13–21. https://doi.org/10.1177/1534765608319323

Hayes, S. C. (2015). *The Act in Context: The Canonical Papers of Steven C. Hayes (1st ed.)*. Routledge. https://doi.org/10.4324/9781315745138.

Kalaitzaki, A., Theodoratou, M., Tsouvelas, G., Tamiolaki, A., & Konstantakopoulos, G. (2024). Coping profiles and their association with vicarious post-traumatic growth among nurses during the three waves of the COVID-19 pandemic. *Journal of Clinical Nursing*. https://doi.org/10.1111/jocn.16988

Kase, S., Waldman, E., & Weintraub, A. (2018). A cross-sectional pilot study of compassion fatigue (CF), burnout (BO), and compassion satisfaction (CS) in pediatric palliative care (PPC) providers (S704). *Journal of Pain and Symptom Management, 55*(2), 657–658. https://doi.org/10.1016/j.jpainsymman.2017.12.358

Kleespies, P. M., Van Orden, K. A., Bongar, B., Bridgeman, D., Bufka, L. F., Galper, D. I., Hillbrand, M., & Yufit, R. I. (2011). Psychologist suicide: Incidence, impact, and suggestions for prevention, intervention, and postvention. *Professional Psychology, Research and Practice, 42*(3), 244–251. https://doi.org/10.1037/a0022805

Levine, P. (2015). *Trauma and memory: Brain and body in a search for the living past: A practical guide for understanding and working with traumatic memory*. North Atlantic Books.

McCann, I. L., & Pearlman, L. A. (1990). Vicarious traumatization: A framework for understanding the psychological effects of working with victims. *Journal of Traumatic Stress, 3*(1), 131–149. https://doi.org/10.1007/bf00975140

McCormack, L., Hagger, M. S., & Joseph, S. (2011). Vicarious growth in wives of Vietnam veterans. *Journal of Humanistic Psychology, 51*(3), 273–290. https://doi.org/10.1177/0022167810377506

Moreau, M. J. (1979). A structural approach to social work practice. *Canadian Journal of Social Work Education, 5*(1), 78–94.

Newell, J. M., & Macneil, G. (2010). Professional burnout, vicarious trauma, secondary traumatic stress, and compassion fatigue. *Best Practices in Mental Health, 6*(2), 57–68.

Ogden, P., & Fisher, J. (2015). *Sensorimotor psychotherapy: Interventions for trauma and attachment (Csm)*. W. W. Norton & Company.

Oshri, A., Cui, Z., Owens, M. M., Carvalho, C. A., & Sweet, L. (2022). Low-to-moderate level of perceived stress strengthens working memory: Testing the hormesis. *Neuropsychologia.*

Puvimanasinghe, T., Denson, L. A., Augoustinos, M., & Somasundaram, D. (2015). Vicarious resilience and vicarious traumatisation: Experiences of working with refugees and asylum seekers in South Australia. *Transcultural Psychiatry, 52*(6), 743–765. https://doi.org/10.1177/1363461515577289

Selye, H. (1976). Stress and physical activity. *McGill Journal of Education, 11*(1), 3.

Siegel, D. J. (2011). *Mindsight: The new science of personal transformation*. Bantam.

Szabo, S., Tache, Y., & Somogyi, A. (2012). The legacy of Hans Selye and the origins of stress research: A retrospective 75 years after his landmark brief "Letter" to the Editor# of Nature. *Stress, 15*(5), 472–478. https://doi.org/10.3109/10253890.2012.710919

Tedeschi, R., & Calhoun, L. G. (1995). *Trauma and transformation: Growing in the aftermath of suffering*. Sage.

Tedeschi, R. G., Shakespeare-Finch, J., Calhoun, L. G., & Taku, K. (2018). *Posttraumatic growth: Theory, research, and applications*. Routledge.

Van Der Kolk, B. A. (2014). *The body keeps the score: Brain, mind, and body in the healing of trauma*. Penguin Publishing Group.

Van Der Kolk, B. A., McFarlane, A. C., & Weisth, L. (1996). *Traumatic stress: The effects of overwhelming experience on mind, body, and society*. Guilford Press.

Whitt-Woosley, A., & Sprang, G. (2023). Analysis of implementation drivers of secondary traumatic stress (STS)-informed organizational change and experiences of STS and burnout in child welfare. *Child Abuse & Neglect, 141*, 106194. https://doi.org/10.1016/j.chiabu.2023.106194

Chapter 6
Integrating Culture into Trauma-Informed Care

1 What Is Culture?

Since American cultural *anthropologists* Kroeber and Kluckhohn (1952) identified the definitions of culture, its complexity and integration into healthcare continue to spark debate. For this chapter, culture refers to the shared understanding of a group's beliefs, values, practices, social behaviors, norms, habits, and systems for navigating the world. Culture influences communication, affecting the significance of physical gestures and the rules governing relationships. While many different cultural groups exist, the two main groups referenced in this chapter are based on individualistic values that shape thinking patterns, processes, and behaviors. The other group comprises non-Western cultures that value interdependence (collectivist/communal culture). However, it is key to remember that culture is fluid and constantly evolves with time, place, and the nature of interactions, education, and population movement.

In social work trauma intervention, it is crucial to consider culture in its temporality and non-static nature. In other words, working with survivors requires paying attention to the survivors' current cultural values and processes that they use to describe experiences and emotions. Your client's current values determine their motivation, capacity, and opportunity to draw on their relationships in healing. Thus, evaluating the values that determine group interaction and support is vital. Cultural groups identify these processes, defining the level of shared experiences (ingroup and outgroup). Ingroup membership is determined by the closeness of the relationship, such as immediate family or those within a close environment, such as parents, siblings, and institutions that are part of the person's or family's life.[1] The outgroup determines external membership, such as tribal members, ethnic group

[1] Family is variously defined. For example, in East Africa, ethnic and tribal groups are considered family.

N. Bokore, *Trauma-Informed, Culturally Based Intervention*,
https://doi.org/10.1007/978-3-032-06835-4_6

members, or communities created during forced migration or resettlement. In trauma care, these identifications define the client–social worker relationships and the emotional connections needed for providing trauma care.

2 What Is Trauma-Informed, Culturally Based Intervention (TICBI)?

In many parts of the world, some health and mental health issues, including trauma response, are stigmatized (see Hailemariam et al., 2016; Jenkins et al., 2010; Kpobi & Swartz, 2018; Tekola et al., 2021). The TICBI approach (detailed in Chaps. 7–10) allows social workers and other professionals involved in trauma care to provide services that include acknowledging the individual's history of displacement, cultural values, beliefs, and beneficial health practices they used in the past. It is rooted in the intervention process that begins with developing a good collaborative working relationship and respecting survivors' self-determination. Providing TICBI is about acknowledging that sometimes individuals and families dealing with mental health and physical illness are bound by cultural taboos, forcing them to hide their symptoms from those they consider as others (outside the family). These cultural barriers are not limited to continental Africa, as people worldwide also experience them. It is about understanding how newcomer clients may seek help only when their symptoms become unbearable or affect their daily life, or when they are mandated by another agency involved in their support system.

Integrating culture in trauma care begins with paying attention to your client's underlying stories and ambivalence about the need for services. Failing to address clients' ambivalence due to fear of offending them can exacerbate service gaps. TICBI assessment (see Chaps. 7 and 8) tackles these issues and what is left unsaid by clients, who might not share the full extent of their pressing problems. Thus, TICBI is about tuning into survivors' stories using the empathic listening skills learned in graduate school. This skill is essential when working with survivors from rural areas facing multiple barriers, such as gaps in education or language skills. In TICBI, tuning-in skills enhance safety and trust while promoting client collaboration.

Understanding the process of cultural communication, including verbal and nonverbal messages, helps reduce tension and address cultural rules and healthcare traditions, especially when you and your client are still strangers. Providing TICBI involves engaging in conversations about traditional healing methods such as herbal remedies, spiritual chants, religious prayers, or dietary restrictions and changes, rather than avoiding or dismissing them. These conversations strengthen the collaborative efforts necessary for change talk, an essential part of trauma care. It is crucial to remember that traditional healing practices are standard in many non-Western cultures, including Africa, particularly where Western-trained health professionals are unavailable. In many parts of the world, traditional healers even serve as primary healthcare providers, filling the healthcare gap (Kale, 1995; Okello & Musisi, 2015; Pillay, 2002). Given the increase in human displacement from war and environmental issues, along with resettlement in Western countries becoming the norm, it is important to understand clients' perceptions of conventional or Westernized practices and their familiarity and trust in traditional healing practices for mental and physical health (Daar et al., 2014; Jenkins et al., 2010; Lund & Flisher, 2009; Maluleka, 2020; Mokgobi, 2014; Ndung'u, 2010; Reuter et al., 2016; Shannon et al., 2015; Tekola et al., 2021; Workneh et al., 2018).

In TICBI, awareness of cultural practices and perspectives is even more critical during psychoeducation. Cultural understanding allows for sharing meaningful information with clients, enhancing their self-understanding and motivation for healing. Cultural communication helps social workers adapt explanations of mental and physical health. For example, some practitioners suggest integrating African drumming and community mentoring as part of psychoeducation and information sharing (Abbo, 2011; Elswick et al., 2021; Im et al., 2018). One notable example is a community-based intervention for traumatized Somali youth in urban Kenya that successfully incorporated cultural values and individual interpretations of experiences into trauma-informed care done in partnership with a community-based organization, community leaders, and mental health workers to share knowledge and its promising results (Im et al., 2018).

Providing TICBI to individuals, their families, and communities requires a collaborative approach that engages them using colloquial terms and everyday language to identify illnesses. The strength of the intervention and collaborative work comes from relying on more than Western-based pathological explanations, often creating a disconnect between clients and social workers. In trauma care, paying attention to what clients share or do not share, along with asking questions as needed, helps collaboratively identify differences, personal views, and ethnocultural factors (Abbo et al., 2008; Abbo, 2011; Daar et al., 2014; Elswick et al., 2021; Patel, 2011).

Providing TICBI in social work involves using an integrated approach that coordinates traditional practices with the explanatory function of Western theories to explore issues affecting clients and plan suitable interventions. Being open and flexible by discussing culturally based interventions such as traditional practices (herbs, supplements, meditation, acupuncture, massage, movement, and nutrition) normalize the intervention process, which is vital to trauma care (Abbo, 2011; Burns & Tomita, 2014; Daar et al., 2014; Patel, 2011).

3 Paying Attention to the Perspectives of Survivors

As aforementioned, culture influences how people express their feelings and thoughts, as well as their behavior, which, in turn, impacts their overall well-being. Western medicine and intervention methods rely on expert knowledge, often silencing those who look and sound different from the typical Western population.

Humans are social beings, and despite variations in their cultures, they are connected through the power of their essential social brain and its functions. Therefore, understanding how collectivist and individualistic cultures provide access to social support systems, ensure safe environments, and promote prosperity through connection is worth understanding. For clients from communal cultures, war, conflict, and displacement disconnect some of these functions, increasing their stress due to the fear of communal judgment. Some survivors adopt maladaptive behaviors, such as self-blame for their experiences (e.g., sexual violence). The fear of bringing shame to the community or family takes priority over seeking help by sharing stressful or stigmatizing stories about themselves. In trauma care, it is crucial to consider these perspectives in each client's history to establish a trusting working relationship. This contrasts with care for clients from individualistic cultures, where intervention is based on task-oriented business relationships and emphasizes discussing issues and solutions.

However, in trauma care, practitioners must address the stressful aspects of the communal culture that often emerge after resettlement. For example, families with young children frequently struggle with bi-culturalism as the younger generation adapts to both the new and old cultures. In TICBI, it is essential to understand parents' and children's perspectives and explain the benefits of bi-culturalism while identifying the sources of their stress or self-isolation. When working with first- or second-generation youth, discussing how parents from individualistic and communal cultures perceive cultural adaptation differently is necessary to facilitate brainstorming these differences for a later discussion. For instance, Western parents may view youths' adoption of punk fashion as a regular part of adolescence. However, non-Western parents may be shocked and concerned about how community members might react. They may focus on concealing the change due to fears of community ostracization and stigmatization. Ostracization and condemnation from the community represent significant life stressors for resettled parents of teenagers.

Therefore, they may choose to self-isolate for their protection, opting for self-seclusion from family elders or religious leaders rather than facing exposure. Social workers providing TICBI must understand that this reaction can become dangerous for some families, potentially leading to death (honor killings). Due to their intense fear of cultural change, families may relocate their children to the village they previously escaped, jeopardizing their safety. Unfortunately, service providers often overlook these challenges in school and healthcare settings, which can have future consequences for the children and their parents. This underscores the necessity of cultural consideration when working with survivors, recognizing its significance in defining every aspect of family or group life. Culturally informed trauma care in social work becomes indispensable as global human movement and displacement continue to rise.

4 Trauma Care in a Multicultural Society

Women often face a higher risk of experiencing conflict-related gender-based violence than men, contributing to the prevalence of PTSD and other health or mental health issues. Cultural norms and expectations regarding gender roles can also influence their responses, affecting how each survivor processes, experiences, or interprets traumatic events (Ainamani et al., 2020). Culturally informed social work intervention involves understanding these differences and acknowledging the multicultural nature of a client's experiences, gendered worldviews, and health-seeking behaviors. It aims to minimize barriers to care services and recognizes how a culture can harm or benefit its members. Every culture has harmful practices and beliefs, some of which particularly affect women, especially with the existence of ritualistic behaviors, such as the normalization of female genital mutilation and child marriage. Multiple arguments are given to justify the ritualistic nature of female genital mutilation; for example, some claim it enhances women's emotional health or protects women's innocence while discouraging sexual violence. Other fear-based explanations include "protecting the girl's parents' honour and preventing blame or shame." As a result, this ritualistic practice continues. Unfortunately, even in some diaspora communities in the West, it is practiced behind closed doors.

To provide trauma care, knowledge of cultural differences in views on gender equality, roles, and family responsibilities is essential. For example, in Western cultures, people generally believe there is gender equity and fairness in distributing benefits and responsibilities among individuals of all genders. However, marked differences still exist in the opportunities for career growth for women and men. It is crucial to remember that women in some non-Western cultures still adhere to their gender-specific roles after migration to maintain the peace and harmony of the family and the group, fearing the cultural consequences of not doing so.

4.1 Gender, Culture, and Trauma Care

When providing TICBI, it is essential to understand the gender-based cultural roles and their differences from Western culture without judgment. Women often adhere to these roles with little choice due to their financial situations, fear, or the need for safety. Creating connections and minimizing the fear of isolation are vital components of TICBI. This approach begins with exploring opportunities for alternatives that align with clients' cultural beliefs. In TICBI, it is crucial to pay attention to gender conformity and its impact on female survivors, such as the fear of being disowned, the risk of honor killing, or other forms of punishment in trauma care. The collective fear of shame and risk is part of human nature and is not limited to women.

In his 1956 study of hypothetical visual perception, social psychologist Solomon Asch explored how persuasion and social pressure affected "individuality versus conformity in a society" (quoted in Gleitman et al., 1997, p. 984). Asch's test aligned with Louise Cozzolino's (2010, p. 177) explanation of the development and function of "the social brain," which informs the toleration and acceptance of cultural gender-based roles, even if they limit women's choices or pressure them to conform for the sake of safety and keeping the family and community harmony intact rather than pursuing individuality or independence. Gender-based conformity has been imposed on women's socialization for centuries, particularly in patriarchal societies where reinforcing specific restrictions benefits the continuity of gender-based power differences.

In the patriarchal culture of some communities, the power of cultural beliefs, particularly when linked to religious and socially constructed norms and expectations, hinders women's ability to adapt to new cultures and live authentically. Moving away from these deeply ingrained norms, which often limit women's choices—such as leaving abusive marital relationships and restricting their self-expression, including their choice of fashion attire—can lead, especially for the second generation, to a disconnect between their true selves and the personas they often present in public, affecting their relationships and overall well-being.

I remember a childhood friend who hid her musical talent and spent her life as a stay-at-home mom and wife, raising her children. There is nothing wrong with choosing to be a stay-at-home mother. The issue arises when it leads to health risks, whether through denying one's self-expression or living with the fear of bringing shame to the family. In my culture, singing is associated with a particular class and an unfavorable lifestyle. Therefore, my friend only sang to her children during girls' nights out or when singing lullabies to other children. Early socialization and adherence to cultural rules and roles limit individual choices and impose conformity, contributing to stress and ultimately affecting an individual's health.

4.2 Culture and Socialization

As discussed above, culture influences gender roles, class structures, health practices, and social interaction. While a culture's harmful and ritualistic practices might persist, beneficial practices and resources also exist. It is vital to understand that culture is part of socialization, which begins early in life and continues with each life experience. In culturally informed trauma care, exploring the cultural rules embedded in a client's life is needed to understand their cognitive processes. Attachment theorist Dr. Allan Schore (2003a, 2003b) describes how early relationships with caregivers determine the human social brain. It is part of the human growth system where the child's caregivers can directly influence the neural wiring and brain circuits responsible for the child's developing brain, determining their later adult life, including how they manage stress and the adherence to rules or roles provided by culture. Consequently, culture plays a role in social relationships and cognitive evaluation, including how individuals perceive or process information based on their values, beliefs, and attitudes. In TICBI, these values and belief systems are considered, and their impact is assessed as part of the individual's life cycle. In TICBI, paying attention to intercultural sensitivity is fundamental, as it is the foundation of both the perception of trauma and the healing process.

Cultural understanding minimizes service barriers, moving survivors from a hyper- or hypoarousal state to a parasympathetic response where they can use their cognitive skills and culture to change (Siegel, 2010). Attuned listening helps social workers adjust cultural communication to resonate with their clients and offer a safe space where they "feel felt" (Siegel, 2010, p. 29). This meaningful understanding reduces clients' stress and widens their window of tolerance. It also helps the social worker see the client's inner world, strengthening the working relationship and promoting change. In TICBI, this process helps understand your client's resilience, perspectives, and life experiences.

Social work values (see Carl Rogers's 1951 Client-centred Therapy) dictate recognizing the complexity of the intersecting social and cultural identities with which clients struggle, as these affect their trauma responses. During TICBI, demonstrating unconditional positive regard toward clients reduces their fear and stress, facilitating the provision of necessary care. Additionally, this professional perspective allows social workers to be flexible, learn and apply new knowledge, engage in periodic self-reflection, practice self-awareness, and prioritize relationships where culturally attuned, informed communication takes center stage.

4.3 Cultural Expressions of Illness

In the current practice setting in Canada, it is not surprising to hear African, Latin American, and Asian clients talking about the consultation they had with traditional healers such as shamans, elders, sheikhs, curanderos, healers, and sangomas. Social workers guided by the principle of "the person-in-their-environment perspective" are often open to hearing their clients' stories and collaborating with them to provide culturally appropriate mental health services that promote well-being.

Until recently, in trauma care, social work was solely focused on the Western-based healthcare system, resisting or discounting knowledge that is considered Other (Bulhan, 1985). The practice continued long after the traditional healers were recognized by the World Health Organization (1978) as gifted people providing health services to their communities using their ancestral knowledge about plants, animals, minerals, and substance use. As we evolved professionally and adopted the value of Indigenous knowledge and holistic practices, traditional healers have gained recognition in many disciplines, including social work. Traditional healers' accumulated knowledge helps them understand the causes of illness, including trauma. This cultural lens allows practitioners the flexibility to pay attention to various perspectives on health, the value of traditional healers, and the roles they play in providing care. Since limited Western-based biomedicine was introduced in many Global South countries during colonialism, each Indigenous group adopted their own cultural and ritualistic allopathic healing methods that are based on their meaning-making process. It is a long-standing health practice familiar to those who believe in it, empowering them to attend to their health. Various authors have explored how people in different parts of the world choose to or have confidence in seeking Indigenous or traditional healing practices before turning to biomedical care. Although the literature is limited, I explore several traditional healing systems that should be considered when providing trauma-informed, culturally based intervention (TICBI) as the first point of care below. However, it is essential to understand the variations in healing practices within these systems.

The expression of trauma differs from culture to culture. Survivors with similar experiences may react differently based on socially informed and traditionally shared concepts of illness and healing. These differences are seen in social work clinical practices, and it is essential in trauma care to be aware of them when these traditional expressions of illnesses are linked to your client's main reason for seeking services. Consequently, paying attention to individual or family reactions, even if perceived as odd, is vital in providing TICBI.

It is also crucial to note that culturally specific descriptions of illness are not limited to particular ethnic groups but are widely used by those from non-Western nations. For example, in many regions of Africa, traditional belief systems explain the causes of mental illness and other health issues as the result of bewitchment. Traditional healers are consulted as the experts in removing the bewitchment and restoring health. According to Moodley and Oulanova (2011), African traditional healers use therapeutic methods they classify as "actions" and "objects" that include a diagnostic process using coffee cup readings or tarot cards, along with many other practices. Some healing actions are "rubbing oils, administering potions, distributing amulets, and performing healing ceremonies, usually at sacred places bearing a mystical-religious meaning" (p. 376). The healing objective of these practices is to understand or determine the course of treatment.

Caribbean and Latin Americans use the *Galphimia glauca* CAV. (golden thryallis) as a homeopathic medicine for *ataques de nervios* (nervous breakdown), which resembles panic attacks. Koro, an illness in many Asian countries, is an anxiety-creating psychiatric disorder caused by psychosexual conflicts, personality, and cultural beliefs. It entails delusional thoughts where men believe that their genitals are shrinking or retracting into their abdomen. It causes fear, terror, and even death. In recent years, Koro has been treated with Western drugs such as anxiolytics, antidepressants, sedatives, and antipsychotic medication. Indigenous practices recommend overcoming stress responses that are causing the delusions or treatments such as exorcism and herbal therapies (Moodley & Oulanova, 2011).

In TICBI, the therapist's awareness of the issues and the survivor's adaptive responses, including those related to cultural learning, such as the expression and belief system of specific illness syndromes, are fundamental to the talk therapy process. The syndromes and explanations of illness mentioned above are just a few examples of cultural expressions of ailment. It is important to remember that even though some syndromes and behaviors may seem bizarre to those from Western cultures, they exist for those who believe in them. Therefore, in social work, these syndromes and the belief in them are considered, acknowledging the diverse ways cultural contexts shape health experiences, normal and abnormal behaviors, and well-being.

5 Valuing Indigenous Practices in Trauma Care

5.1 *Africa*

Growing up, hearing about illness and seeking the expertise of traditional healers for care was a practice that was normalized in my community. As an 11-year-old, I came close to death from meningitis. Extraordinary traditional healers used herbal medicine, and religious leaders prayed, read religious scriptures, and applied holy water, which somehow brought me back to health. I was in a small town far away from a large city where there was no modern healthcare provider, and without those herbal medicines, I would have suffered the consequences, perhaps even died. The power of curing meningitis by herbal medicine and prayer may not make sense for many Western-trained practitioners, but it worked for me. We could debate whether the firm belief of those caring for me helped my body/brain heal itself (the placebo effect) or whether centuries-old medicinal properties within those herbs healed me.

I am not generalizing my experience to the 54 African countries where traditional, holistic healing methods are used. Even though they are different because the cultures in which they are based differ, they have in common the use of traditional healers who have provided services when there is no other option. As a result, many in the African diaspora still use herbal medicine as a trusted method for minor health issues, delaying seeking help from biomedical practitioners. Many still believe that traditional health practice provides holistic health, which differs from the Western-based healthcare system that focuses on psycho-pathologizing health-related issues, making belief in alternative remedies unacceptable.

A belief in traditional practices has solid social support in many parts of Africa and is encouraged by religious, spiritual, and tribal elders in countries throughout the continent (Alem et al., 2008; Mokgobi, 2014; Ndung'u, 2010; Okello & Musisi, 2015). Some of the remedies involve elders encouraging those with ill health to believe in their connection with ancestors and the land to promote well-being because they believe in the impact of generational disconnection causing ill health (Mokgobi, 2014; Sankoh et al., 2018; Van Der Watt et al., 2018; Van Der Zeijst et al., 2021; Workneh et al., 2018). I have seen traditional healers providing survivors of trauma faith-based self-regulation, benefiting from increasing self-awareness, self-care, and community nurture, and identity formation after displacement, with positive results and health benefits, including promoting post-traumatic growth (see Chap. 4).

Therefore, valuing the role of traditional healers in trauma care and integrating cross-cultural understanding and meaning-making processes in trauma care are essential aspects of providing TICBI. It is about shifting the conversation with clients of African descent from deficit-based care to hope-generating, holistic trauma care. Even though evaluating the effectiveness of integrating traditional and biomedical care is not yet possible, the practice helps service providers combine evidence-based practices with culturally relevant knowledge (Abbo, 2011; Ben-Ezra et al., 2020).

5.2 Latin America

Similarly, within the 33 Latin American countries, varied Indigenous methods connect the mind, body, and spirit. Throughout Latin America, Indigenous practice also connects the earth, environment, community, and self as the center of providing holistic practices supported by primary healthcare providers in Western countries (Lucana & Elfers, 2020; Pereira & Andrade, 2018). For example, the ayahuasca plant is now discussed by trauma experts like Dr. Gabor Maté and Dr. van der Kolk as a plant-based psychedelic healer, matching the idea of its use for centuries in many Latin American countries (Perkins et al., 2021; Politi et al., 2021; Weiss et al., 2021).

The benefits of these ancient practices and their integration into modern medicine were explored in a study by Perkins et al. (2021) of ayahuasca[2] drinkers in 40 countries, whose health outcomes changed positively. The study also recognized the need to combine the practice of the spiritual counselor/leader with that of the psychotherapist. In Latin America, consuming ayahuasca is associated with ceremonial gatherings where significant preparation is undertaken to ensure positive healing outcomes of participants (Ghelman et al., 2021; Lucana & Elfers, 2020; Perkins et al., 2021; Politi et al., 2021; Weiss et al., 2021).

5.3 Asia

Chinese medicine is gaining popularity in Western health practices. Before gaining popularity in the West, Chinese conventional medicine had been practiced for millennia, appearing in medical texts such as Huangdi's *Internal Classics*, published c. 300 BCE. Its traditional mental health counseling methods are similar to Western-based psychotherapy (Gu & Pei, 2017). Other Chinese healing methods have also made their way into Western practices, such as the now-common health practice of acupuncture to improve healing (Gielen et al., 2013; Gu & Pei, 2017).

Yin-Qi and Yang-Qi are two Asian healthcare approaches associated with feminine and masculine energies. Some mental health therapists now employ a balance

[2] Ayahuasca is used for spiritual, personal cleansing, and emotional growth. Other mental health benefits include a cure for depression, anxiety, mood disorders, and drug addiction. Ayahuasca is also credited for its anti-inflammatory and antimicrobial properties.

between the two to promote well-being (Hadad, 2013, p. 67). These traditional healing practices are also utilized to explain how trauma and other illnesses impact well-being. For example, in traditional Chinese medicine, the connection between the physical, including internal organs, and emotions is described by saying, "consistent rage impairs the liver, sorrow prevails over anger; elation prevails over fear; anxiety impairs the spleen; grief impairs the lungs, and joy prevails over melancholy" (Gielen et al., 2013, p. 183).

Indian Hindu texts describe how the human body consists of three essential elements (*gunas*), wind (*vata*), bile (*pitta*), and phlegm (*kapha*) and are often used by holistic health practitioners (Gielen et al., 2013, p. 197). Indian gurus have explained the body/brain's connection with the environment and society for centuries. An example used in Hindu teaching is how health issues result from not living up to the standard dictated by collectivist and communal social norms, or related to misdeeds in a past life. Another example is shame associated with breaking social norms, which is considered a source of ill health. Traditional interventions are used to recover from these challenges, including therapeutic intervention informed by Ayurveda, Unani, Sahaja, Shamanism, Sidda, homeopathy, naturopathy, fortune tellers, and astrologers. These traditional Asian healing practices are beginning to be incorporated into mainstream practice in India and other Asian countries (Gielen et al., 2013).

These examples demonstrate the awareness of Indigenous practices required in trauma care. The client's traditional practices or belief systems are crucial in TICBI and other social work interventions. As culture shapes the brain, thought processes, values, and behaviors, and defines the collaborative process during trauma care, it is essential to reflect on rich historical healing practices alongside modern health services to provide person-centered care.

6 Culture and the Neurobiological Legacy of Trauma

Humans are a cultural species and respond to traumatic experiences according to their culture, which shapes their functional brain responses and adaptation (Ambady & Bharucha, 2009). A client's ethnocentric worldview often shapes their mind in the aftermath of trauma, and the varied individual responses of clients reflect cultural adaptation and how it determines their reactions. This does not mean a change in the functional structure of the brain response; it means an event changes the direction and, therefore, the connectivity of neurons determining body/brain responses based on a cultural or spiritual interpretation of the event. Cultural and spiritual beliefs become a protective cushion against the overwhelming neurobiological and psychological effects of dysregulated autonomic responses and their devastating impact on survivors. These teachings become adaptive strategies that implicitly function as the body's automatic survival defense system (Ogden & Fisher, 2015; Ogden et al., 2006). Cultural teaching makes a difference in neural wiring because of genetic endowment, brain plasticity, intercultural relationships, and environmental input. Cultural teachings create a sensory adaptation that the body and mind use to sustain life. When facing the danger of war and displacement, these survival resources keep a survivor's mind from engaging in reactive responses that further endanger their life.

Our culturally structured minds affect how we think, feel about, and react to trauma. While some survivors are impacted by the slightest hint of danger, defining their overwhelming emotional and physical reaction to others, their adaptive responses change their sensations and actions about the event (Ogden & Fisher, 2015; Ogden et al., 2006). The human mind, shaped by culture, is a fantastic structure that determines human cognition, memories, and behavioral reactions. It establishes the mind (thinking and feeling) and the meaning given to past experiences, changing the brain's function, which is the organ of the mind (Siegel, 2007). Therefore, the quality of the mind depends on the function and reaction of the brain.

These cultural and neurobiological structures and responses are seen in the human emotional system, based on the perception and interpretation of events in the ventral mPFC area of neural activities whose role is to represent context, events, and responses (Chiao, 2009). Based on cultural learning, variations are also seen in other areas, including how empathy, impulse control, emotion, and decision-making are processed by the anterior cingulate cortex (ACC). These variations define the trauma response of individuals, regardless of culture, and the variations are also present in the brain process in self-expression and areas involved in close relationships, including family and friends (Ames & Fiske, 2010; Bernardi et al., 2019).

For individuals from individualistic cultures, brain activity is observed in the ventral medial prefrontal cortex (VmPFC) and posterior cingulate cortex (pCC) when answering questions related to emotional memories of closeness to others. However, when individuals from collectivist cultures are asked the same question (emotional memories related to closeness to others), contemplating the potential social risks that could cause stigma and affect self-esteem leads to a similar activation of the VmPFC and pCC (Chiao 2009). Understanding these differences and how culture shapes the questions asked in therapy and activating the brain areas helps elucidate the survivor's thought process and the necessary intervention. However, these thought processes are not set in stone; they can change because culture is temporal. Therefore, changes occur with interpreting experiences based on new learning that determines brain activation (Han, 2015).

Thus, TICBI requires a reflective conceptualization of the presenting issues and a culturally informed thought process. As a result, intervention requires carefully collected data during the bio-psychosocial-spiritual assessment (see Chaps. 7 and 8). Since each life experience brings new synaptic connections, every trauma care session is seen as helping survivors create new neural connections that become the inception of new memories, growth, and change.

7 The Legacy of Culture and Stress in Trauma Care

In trauma care, it is essential to recognize that "trauma affects the entire human organism—body, brain and mind" (van der Kolk, 2014, p. 53), which is built around socially constructed and culturally reinforced ways of thinking. Therefore, exploring how culturally learned stress responses affect the body, brain, and mind and how they define trauma response is fundamental. For example, while some clients may consider war and displacement stress as symptoms of their temporary situation, others, again based on their culture and belief system, might see it as divine punishment for past deeds requiring change through good deeds, prayers, and asking forgiveness from a higher power. During TICBI, discussing belief systems and cultural or ethno-racial beliefs is part of planning TICBI.

The latest global community health movement and belief in the importance of Indigenous knowledge have created opportunities for social workers to raise awareness of service gaps in trauma care, resulting from a lack of understanding or overlooking traditional and cultural aspects of stress response (Bodini, 2023). This global movement has highlighted the long history of colonized healthcare and mental health discourses that did not consider the stress and trauma responses of multicultural, multi-ethnic, and multi-religious communities, as well as the benefits of their traditional healing practices as credible and complementary resources. Although anti-oppression practices in trauma care have increased awareness, Western healthcare systems, viewed as superior to traditional care systems, continue to be taught in medical schools without the cultural integration of care. TICBI invites service providers to acknowledge these gaps and consider varied trauma responses and healing practices. It creates a space for the needs of survivors from non-Western cultures and challenges the status quo rooted in Eurocentric training in the provision of trauma care services.

8 Providing Culturally Based Trauma-Informed Intervention (TICBI)

TICBI follows the person-centered approach, which prioritizes survivors' needs and uses cultural communication to understand sources of trauma and triggers. One of the guiding principles of a person-centered approach is to focus on the uniqueness of the individual and their trauma responses. It helps avoid the one-size-fits-all Western approach currently used in the healthcare system. It focuses on survivors' worldviews, belief systems, subjective experiences, and how culture determines their health-seeking behaviors. Thus, providing TICBI includes familiarizing yourself with the communities you serve and asking yourself the following questions:

- What is the demography of the area I am serving?
- What ethnicities and cultures do I need to be aware of?
- What are the perceptions of health and systems of care I need to consider?
- What are clients' health-seeking behaviors, and what are their preferred treatments?

TICBI is a holistic approach to servicing multicultural/multi-ethnic/multi-religious communities (see Chaps. 7 and 8). TICBI allows service delivery rooted in social work's egalitarian principles of equal rights, equal access to services, and a practice based on anti-oppression. When you work with clients as their social worker, and you begin to doubt your client's descriptions of their trauma and mental health issues, ask yourself these questions:

- Am I misinterpreting my client based on what I know about symptom descriptions (DSM or ICD-10-CA, which are used worldwide, and the Canadian Classification of Health Interventions)?
- Am I missing information lost in cultural miscommunication, causing my misinterpretation or misunderstanding, and how do I change that?

TICBI begins with assessing your client's needs, values, perspectives, and experiences, including their pre-conflict and post-conflict relationships, health issues, and culturally informed practices. This practice perspective helps understand the presenting matters and your client's support system that shields them from experiencing the adverse effects of war, displacement, and resettlement stress by regulating their emotions and promoting post-traumatic growth (Schwartz & Shrira, 2019; Siegel, 2011). Use the questions and actions shown in Table 6.1 to learn more about your client's cultural values and plan an appropriate intervention.

In TICBI, the above exploratory questions rely on open and attuned communication within an empathetic and compassionate relationship (Ager, 1999; Porges & Dana, 2018; Siegel, 2011). This approach enables social workers to effectively conduct TICBI assessments and plan interventions while acknowledging and respecting the survivor's values and belief systems that can be integrated to tailor the intervention. TICBI focuses on guiding clients from a fight-or-flight response (sympathetic nervous system activation) or freeze response (dorsal vagal system activation) to a rest-and-digest state (ventral vagal state) (Cozolino, 2010; Porges & Dana, 2020). Therefore, the intervention process centers on co-regulation to promote well-being. It is grounded in collaborative interaction and knowledge exchange based on a calm, attuned presence where the client's autonomic ventral state is achieved through the benefits of the therapeutic space. It serves as a professional reminder that collaboration for change is possible when culturally informed, respectful communication built on curiosity, care, and compassion is initiated.

Table 6.1 Identifying clients' cultural values

Questions	Actions
Is this your first experience seeking community, health, and mental health services?	If yes, explain your services If no, ask what brought them to seek this service, how they learned about it, and their expectations
In your culture, when similar issues arise, where do you go to seek help?	When traditional practices are mentioned, ask related questions out of curiosity and a desire to learn
What is the most crucial issue contributing to the matter, and if it happened in the past, how did you resolve it?	Listen to learn and take notes for future discussion. Ask additional questions if you notice the use of culturally based holistic health practices
Is this practice the same for your family and community, and how did it contribute to their well-being?	Take notes for future discussion about past healing or recovery practices for similar issues that brought the client to you
How are emotions and other health-related issues discussed in your family or community?	Take notes for future discussion about culturally informed and preferred healing practices
How significant are spirituality, religion, and traditional/cultural values?	Listen and take notes to include in healing and well-being discussions and psychoeducation
In your culture or family tradition, what are the roles of elders, religious leaders, healers, men, women, and children?	Listen and take notes for a later discussion about talk therapy/talking cure and healing practices

9 TICBI Challenges

One of the fundamental values of social work is acknowledging the diversity of knowledge and differences in worldviews. It follows the client-centered approach, where the social worker provides TICBI and begins their services by establishing respectful relationships based on recognition, human connection, and constant self-reflection. TICBI is about providing an environment of safety where energy and thoughts are exchanged (mirroring and co-regulation), and the social worker presents energy throughout the process, saying, "I see you, I hear you, and I am here to help."

Graduate social work students are often taught cultural competency and sensitivity skills. Cultural competency is recognizing clients' cultural beliefs and values and how they influence their implicit actions. On the other hand, cultural sensitivity is defined as recognizing, understanding, and reacting appropriately to the opinions and behaviors of groups seen as Other. However, these two skills limit and overlook the diversity of cultures and subcultures. Cultural understanding begins in working with individual clients and, as mentioned in this chapter, exploring the many factors that define culture with them.

Culturally informed intervention is necessary in providing TICBI as it shapes the social worker–client interaction and the change process. It is based on an intersectional understanding of trauma and its intergenerational effects on past wounds and on collaboratively discussing healing methods. Therefore, the challenge for TICBI is understanding the role of culture in the unique expression of trauma and healing practices that make meaning for survivors. As mentioned, culture, traditional practices, and clients' needs are central to assessment and intervention.

Definitions

Allopathic healing: A system of medical practice that emphasizes illness diagnosis and treatment.

Anterior cingulate cortex (ACC): This unique structure in the forebrain is connected to both the "emotional" limbic system and the "cognitive" prefrontal cortex.

Communal, collectivist culture: A culture that values the group over the individual, generosity over selfishness, harmony over conflict, and meeting group needs over individual needs. Emphasizes the significance of community, collectivity, and collaboration.

Culture: For this book, culture is the shared beliefs, language, norms, values, and symbols used by groups to provide them with a framework of life, values, ethics, and community.

Indigenous practice: Even though it is often associated with North American Indigenous groups, Indigenous practice defines traditional practices, knowledge, skills, and techniques used by Indigenous communities worldwide.

Individualistic culture: A culture focused on rights and concerns, which emphasizes independence, personal goals, and achievements.

Intercultural sensitivity: The awareness of subjective cultural contexts and barriers clients struggle with.

Posterior cingulate cortex (pCC): A brain region strongly implicated in various aspects of cultural and cognitive processes, including the self, social cognition, and emotional regulation. It plays a role in trauma and cultural responses. Disruptions in this area affect connectivity with other brain regions and are linked to PTSD, impacting connections with the prefrontal cortex, amygdala, and hippocampus.

Traditional healers: Elders who know about health, healing, and medicine. They are mainly known in non-Western cultures and Indigenous groups. They provide traditional medicine and teach or mentor the next generation of traditional healers.

Ventromedial prefrontal cortex (vmPFC): The vmPFC is part of the prefrontal cortex linked to the frontal lobe and the bottom of the cerebrum (left and right). It is involved in processing risk and fear. It serves as an integrative hub for processing emotional, sensory, social, memory, and self-related information influenced by cultural values and adapted norms that shape the functioning of the vmPFC. For example, its functions affect how people in some cultures emphasize the impact of collective negative experiences, while others prioritize individual consequences.

Yin-Qi and Yang-Qi (the material basis of blood circulation and functional activities of the body, respectively): The interaction between what is commonly known as Yin and Yang generates Qi. In Taoism, the balance of these two opposites is necessary as they revolve in harmony and promote well-being. It is believed that Yin exists within Yang, and the two are inextricably connected. Yin symbolizes softness, introspectiveness, femininity, and healing, while Yang symbolizes strength, hardness, energy, masculinity, and sometimes aggression.

Discussion Questions

1. Is trauma response universal or culturally constructed?
2. Why is knowing about traditional expressions of illnesses and healing practices important?
3. Discuss the importance of self-awareness in providing TICBI.
4. Explain the differences between cultural competency and cultural sensitivity. Why is the awareness of your client's needs based on culture important in providing trauma care?
5. How do your cultural background, educational training, and experiences influence your attitudes, values, and biases toward traditional healing practices? Are adjustments necessary?
6. Summarize what you learned from this chapter.
7. If the client you are working with is experiencing one of the cultural illnesses described in this chapter, how will you address it, and what will you include in your intervention planning?
8. Define the following terms and discuss their relevance to social work practice: multiculturalism, intercultural sensitivity, ethnocentric worldview.Competing InterestsThe author has no conflicts of interest to declare that are relevant to the content of this chapter.

References

Abbo, C. (2011). Profiles and outcome of traditional healing practices for severe mental illnesses in two districts of eastern Uganda. *Global Health Action/Global Health Action. Supplement, 4*(1), 7117. https://doi.org/10.3402/gha.v4i0.7117

Abbo, C., Ekblad, S., Waako, P., Okello, E., Muhwezi, W., & Musisi, S. (2008). Psychological distress and associated factors among the attendees of traditional healing practices in Jinja and Iganga districts, eastern Uganda: A cross-sectional study. *International Journal of Mental Health Systems, 2*(1), 16. https://doi.org/10.1186/1752-4458-2-16

Ager, A. (1999). *Refugees: Perspectives on the experience of forced migration*. Pinter.

Ainamani, H. E., Elbert, T., Olema, D. K., & Hecker, T. (2020). Gender differences in response to war-related trauma and posttraumatic stress disorder—A study among the Congolese refugees in Uganda. *BMC Psychiatry, 20*(1). https://doi.org/10.1186/s12888-019-2420-0

Alem, A., Jacobsson, L., & Hanlon, C. (2008). Community-based mental health care in Africa: Mental health workers' views. *World Psychiatry/World Psychiatry, 7*(1), 54–57. https://doi.org/10.1002/j.2051-5545.2008.tb00153.x

Ambady, N., & Bharucha, J. (2009). Culture and the brain. *Current Directions in Psychological Science, 18*(6), 342–345. https://doi.org/10.1111/j.1467-8721.2009.01664.x

Ames, D. L., & Fiske, S. T. (2010). Cultural neuroscience. *Asian Journal of Social Psychology, 13*(2), 72–82. https://doi.org/10.1111/j.1467-839x.2010.01301.x

Ben-Ezra, M., Hyland, P., Karatzias, T., Maercker, A., Hamama-Raz, Y., Lavenda, O., Mahat-Shamir, M., & Shevlin, M. (2020). A cross-country psychiatric screening of ICD-11 disorders specifically associated with stress in Kenya, Nigeria and Ghana. *European Journal of Psychotraumatology, 11*(1).

Bernardi, J., Engelbrecht, A., & Jobson, L. (2019). The impact of culture on cognitive appraisals: Implications for the development, maintenance, and treatment of posttraumatic stress disorder. *Clinical Psychologist, 23*(2), 91–102. https://doi.org/10.1111/cp.12161

Bodini, C. (2023). The key role of social movements in protecting the health of people and the planet; Comment on "Ensuring global health equity in a post-pandemic economy". *International Journal of Health Policy and Management*. https://doi.org/10.34172/ijhpm.2023.7757
Bulhan, H. A. (1985). *Frantz fanon and the psychology of oppression*. Plenum Press.
Burns, J. K., & Tomita, A. (2014). Traditional and religious healers in the pathway to care for people with mental disorders in Africa: A systematic review and meta-analysis. *Social Psychiatry and Psychiatric Epidemiology, 50*(6), 867–877. https://doi.org/10.1007/s00127-014-0989-7
Chiao, J. Y. (2009). *Cultural neuroscience: Cultural influences on brain function*. Elsevier.
Cozolino, L. J. (2010). *The neuroscience of psychotherapy: Healing the social brain* (2nd ed.). W. W. Norton & Company.
Daar, A. S., Jacobs, M., Wall, S., Groenewald, J., Eaton, J., Patel, V., Santos, P. D., Kagee, A., Gevers, A., Sunkel, C., Andrews, G., Daniels, I., & Ndetei, D. (2014). Declaration on mental health in Africa: Moving to implementation. *Global Health Action, 7*(1). https://doi.org/10.3402/gha.v7.24589
Dana, D. (2020). *Polyvagal exercises for therapists and clients: 50 client-centered practices*. W. W. Norton & Company.
Elswick, S., Washington, G., Mangrum-Apple, H., Peterson, C., Barnes, E., Pirkey, P., & Watson, J. (2021). Trauma healing club: Utilizing culturally responsive processes in the implementation of an after-school group intervention to address trauma among African refugees. *Journal of Child & Adolescent Trauma, 15*(1), 155–166. https://doi.org/10.1007/s40653-021-00387-5
Ghelman, R., Sibinga, E. M. S., Von Schoen-Angerer, T., Vagedes, J., & Seifert, G. (2021). The São Paulo agreement on integrative pediatrics: A consensus-based document fostering integrative health of children and adolescents globally. *Complementary Therapies in Medicine, 58*, 102677. https://doi.org/10.1016/j.ctim.2021.102677
Gielen, U. P., Moodley, R., & Wu, R. (2013). *Handbook of counseling and psychotherapy in an international context*. Routledge.
Gleitman, H., Rozin, P., & Sabini, J. (1997). Solomon E. Asch (1907–1996). *American Psychologist, 52*(9), 984–985. https://doi.org/10.1037/0003-066x.52.9.984
Gu, S., & Pei, J. (2017). Innovating Chinese herbal medicine: From traditional health practice to scientific drug discovery. *Frontiers in Pharmacology, 8*, 381. https://doi.org/10.3389/fphar.2017.0038
Hadad, M. (2013). *Satisfying life: Positive psychology and personal growth*. Pearson Canada.
Hailemariam, M., Fekadu, A., Selamu, M., Medhin, G., Prince, M., & Hanlon, C. (2016). Equitable access to integrated primary mental healthcare for people with severe mental disorders in Ethiopia: A formative study. *International Journal for Equity in Health, 15*(1). https://doi.org/10.1186/s12939-016-0410-0
Han, S. (2015). Understanding cultural differences in human behavior: A cultural neuroscience approach. *Current Opinion in Behavioral Sciences, 3*, 68–72. https://doi.org/10.1016/j.cobeha.2015.01.013
Im, H., Jettner, J. F., Warsame, A. H., Isse, M. M., Khoury, D., & Ross, A. I. (2018). Trauma-informed psychoeducation for Somali refugee youth in urban Kenya: Effects on PTSD and psychosocial outcomes. *Journal of Child & Adolescent Trauma, 11*(4), 431–441. https://doi.org/10.1007/s40653-017-0200-x
Jenkins, R., Heshmat, A., Loza, N., Siekkonen, I., & Sorour, E. (2010). Mental health policy and development in Egypt—Integrating mental health into health sector reforms 2001-9. *International Journal of Mental Health Systems, 4*(1), 17. https://doi.org/10.1186/1752-4458-4-17
Kale, R. (1995). Traditional healers in South Africa: A parallel health care system: South Africa's health. *British Medical Journal, 6899*(310), 1182–1186.
Kpobi, L., & Swartz, L. (2018). Implications of healing power and positioning for collaboration between formal mental health services and traditional/alternative medicine: The case of Ghana. *Global Health Action, 11*(1), 1445333. https://doi.org/10.1080/16549716.2018.1445333
Kroeber, A. L., & Kluckhohn, C. (1952). *Culture : A critical review of concepts and definitions*. Vintage Books.

Lucana, S., & Elfers, J. (2020). Sacred Medicine: Indigenous healing and mental health. *The Qualitative Report*. https://doi.org/10.46743/2160-3715/2020.4626.

Lund, C., & Flisher, A. J. (2009). A model for community mental health services in South Africa. *Tropical Medicine & International Health, 14*(9), 1040–1047. https://doi.org/10.1111/j.1365-3156.2009.02332.x

Maluleka, R. J. (2020). The status of traditional healing in the Limpopo province of South Africa. *HTS Teologiese Studies/Theological Studies, 76*(4). https://doi.org/10.4102/hts.v76i4.6103

Mokgobi, M. G. (2014). Understanding traditional African healing. *African Journal for Physical Health Education, Recreation, and Dance, 20*(Suppl 2), 24–34.

Moodley, R., & Oulanova, O. (2011). Rhythm of the pulse: South Asian traditional healers and their healing practices in Toronto. *International Journal of Health Promotion and Education, 49*(3), 90–100. https://doi.org/10.1080/14635240.2011.10708215

Ndung'u, N. (2010). Persistence of features of traditional healing in the churches in Africa: The case of the Akurinu churches in Kenya. *Africa Thought and Practice/Thought and Practice, 1*(2). https://doi.org/10.4314/tp.v1i2.51596

Ogden, P., & Fisher, J. (2015). *Sensorimotor psychotherapy: Interventions for trauma and attachment (Csm)*. W. W. Norton & Company.

Ogden, P., Minton, K., & Pain, C. (2006). *Trauma and the body: A sensorimotor approach to psychotherapy* (Norton Series on Interpersonal Neurobiology) (1st ed.). W. W. Norton & Company.

Okello, E., & Musisi, S. (2015). The role of traditional healers in mental health care in Africa. In E. Akyeampong, A. G. Hill, & A. Kleiman (Eds.), *The culture of mental illness and psychiatric practice in Africa* (pp. 249–261). Indiana University Press.

Patel, V. (2011). Traditional healers for mental health care in Africa. *Global Health Action, 4*(1), 7956. https://doi.org/10.3402/gha.v4i0.7956

Pereira, A., & Andrade, D. C. L. (2018). Mental health educational program for primary health care physicians. *Revista Brasileira De Educação Médica, 42*(1), 6–14. https://doi.org/10.1590/1981-52712015v41n4rb20160021

Perkins, D., Schubert, V., Simonová, H., Tófoli, L. F., Bouso, J. C., Horák, M., Galvão-Coelho, N. L., & Sarris, J. (2021). Influence of context and setting on the mental health and well-being outcomes of ayahuasca drinkers: Results of a large international survey. *Frontiers in Pharmacology, 12*. https://doi.org/10.3389/fphar.2021.623979

Pillay, A. L. (2002). Rural and urban South African women's awareness of cancers of the breast and cervix. *Ethnicity and Health, 7*(2), 103–114. https://doi.org/10.1080/1355785022000038588

Politi, M., Tresca, G., Menghini, L., & Ferrante, C. (2021). Beyond the psychoactive effects of ayahuasca: Cultural and pharmacological relevance of its emetic and purging properties. *Planta Medica, 88*(14), 1275–1286. https://doi.org/10.1055/a-1675-3840

Porges, S. W., & Dana, D. (2018). *Clinical applications of the polyvagal theory: The emergence of polyvagal-informed therapies* (Norton series on interpersonal neurobiology). W. W. Norton & Company.

Reuter, P. R., McGinnis, S. M., & Reuter, K. E. (2016). Public health professionals' perceptions of mental health services in Equatorial Guinea, central-West Africa. *Pan African Medical Journal, 25*, 10.11604/pamj.2016.25.236.10220.

Rogers, C. R. (1951). *Client-centered therapy: Its current practice, implications and theory*. Houghton Mifflin.

Sankoh, O., Sevalie, S., & Weston, M. (2018). Mental health in Africa. *The Lancet Global Health, 6*(9), e954–e955. https://doi.org/10.1016/s2214-109x(18)30303-6

Schore, A. N. (2003a). *Affect dysregulation and disorders of the self.* W. W. Norton & Company.

Schore, A. N. (2003b). *Affect regulation and the repair of the self.* W. W. Norton & Company.

Schwartz, E., & Shrira, A. (2019). Social connectedness moderates the relationship between warfare exposure, PTSD symptoms, and health among older adults. *Psychiatry, 82*(2), 158–172. https://doi.org/10.1080/00332747.2018.1534521

Shannon, P. J., Wieling, E., Simmelink-McCleary, J., & Becher, E. (2015). Beyond Stigma: Barriers to Discussing Mental Health in Refugee Populations. *Journal of Loss and Trauma, 20*(3), 281–296. https://doi.org/10.1080/15325024.2014.934629.

Siegel, D. J. (2007). *The mindful brain: Reflection and attunement in the cultivation of well-being*. Norton & Company.

Siegel, D. J. (2010). *The Mindful Therapist: A Clinician's guide to mindsight and Neural Integration (Norton Series on Interpersonal Neurobiology)*. W. W. Norton & Company.

Siegel, D. J. (2011). *Mindsight: The new science of personal transformation*. Bantam.

Tekola, B., Mayston, R., Eshetu, T., Birhane, R., Milkias, B., Hanlon, C., & Fekadu, A. (2021). Understandings of depression among community members and primary healthcare attendees in rural Ethiopia: A qualitative study. *Transcultural Psychiatry, 60*(3), 412–427. https://doi.org/10.1177/13634615211064367

Van Der Kolk, B. A. (2014). *The body keeps the score: Brain, mind, and body in the healing of trauma*. Penguin Publishing Group.

Van Der Watt, A. S., Menze, N., Moxley, K., & Seedat, S. (2018). Profiles of traditional healers and their healing practices in the Eastern Cape province of South Africa. *South African Journal of Psychiatry, 24*. https://doi.org/10.4102/sajpsychiatry.v24i0.1305

Van Der Zeijst, M., Veling, W., Makhathini, E. M., Susser, E., Burns, J. K., Hoek, H. W., & Susser, I. (2021). Ancestral calling, traditional health practitioner training and mental illness: An ethnographic study from rural KwaZulu-Natal, South Africa. *Transcultural Psychiatry, 58*(4), 471–485. https://doi.org/10.1177/1363461520909615

Weiss, B., Miller, J. D., Carter, N. T., & Campbell, W. K. (2021). Examining changes in personality following shamanic ceremonial use of ayahuasca. *Scientific Reports, 11*(1). https://doi.org/10.1038/s41598-021-84746-0

Workneh, T., Emirie, G., Kaba, M., Mekonnen, Y., & Kloos, H. (2018). Perceptions of health and illness among the Konso people of southwestern Ethiopia: Persistence and change. *Journal of Ethnobiology and Ethnomedicine, 14*(1). https://doi.org/10.1186/s13002-018-0214-y

World Health Organization. (1978). *The promotion and development of traditional medicine: Report of a WHO meeting*. https://iris.who.int/bitstream/handle/10665/40995/WHO_TRS_622.pdf?sequence=1

Part IV
Planning TICBI

Chapter 7
Planning an Assessment

1 Planning the Initial Bio-Psychosocial, Structural, and Cultural Assessment

Healing from trauma is possible because the human brain constantly changes based on experience, modifying itself and creating new neural connections, facilitating adjustments through brain plasticity. Eric Kandel's (2006) groundbreaking study explains the human brain's capacity to change brain cells based on new learning and the creation of new memories, altering both the neural connections and the brain structure in the process. Eric Kandel's discovery contributes to the understanding of trauma response and the healing process needed for providing TICBI, which begins with carefully planning the assessment and intervention goals. It helps define both the traumatic experiences and the potential treatment outcomes for survivors based on the three mindsets discussed in this chapter.

Providing TICBI for the first time can be stressful and intimidating, especially for recent graduates beginning their careers in talk therapy. Therefore, seeking support from colleagues and supervisors about the process is a valuable starting point. Another proper preparatory trauma intervention strategy is reading about the latest methods relevant to upcoming sessions. Preparation is an empowering tool for social workers, enhancing their commitment to delivering a service rooted in a client-centered approach.

Service organizations providing TICBI often employ intake workers or receptionists as the first point of contact in identifying service needs. During this process, helpful information is typically gathered, including the referral source (family members, physicians, judges, teachers, religious leaders, and other helping professionals) and other essential details. This information enables the social worker to engage the client effectively, compile additional details, establish a connection, and utilize emotional co-regulation to support the intervention process.

N. Bokore, *Trauma-Informed, Culturally Based Intervention*,
https://doi.org/10.1007/978-3-032-06835-4_7

Using this information, preparing to provide TICBI involves conducting a detailed multidimensional assessment. It is also crucial to be aware of the intersectionality of oppression—race, religion, gender, ability, and sexuality—and how these factors influence the impact of trauma based on demographic identifiers. In TICBI, during the pre-planning phase, considering relevant theories that support the assessment, including the client's bio-psychosocial, cultural, and structural makeup, as well as the issues that led them to seek help, enhances the intervention.

Pre-planning the bio-psychosocial, structural, and cultural assessment (see Chap. 8) requires understanding the client's mindset, needs, and specific issues. In these initial steps, theoretical application allows social workers to explore the precipitating factors of the current problem, the perpetuating or maintaining factors that create stuck points, and the protective factors related to the issues and intervention process (Bogo, 2006). For example, the growth of neuroscience in the past two decades has helped social workers focus on trauma responses and their impacts on the mind, body, and brain, offering opportunities to collaborate with the client and explore the issues they are facing holistically.

Thus, planning a TICBI assessment enables social workers to focus on the biological dimension of trauma, including brain and body responses, while identifying possible internal predispositions that may increase the client's vulnerability to physical and mental illnesses. Understanding the biological process of trauma allows them to explore and discuss with clients triggers and maladaptive responses they might be using to overcome external stress, such as living in a toxic environment (e.g., alcohol, drug, or other substance use).

Planning a psychological assessment in social work often relies on identifying symptoms as outlined in the DSM-5 Clinician-Administered PTSD Scale (CAPS-5) or on diagnosing acute stress disorder or an adjustment disorder (Edwards et al., 2023). While these are significant resources for TICBI assessment, it is also essential to incorporate the identification of brain–body responses to trauma to evaluate the causes of a client's hyperarousal or hypoarousal, anxiety or hyperreactivity, exhaustion, confusion, sadness, and numbness. These responses are often observed in trauma survivors exhibiting a survival mindset or, in certain situations (during a crisis), in those with a therapeutic or active, engaged mindset. For assessing the social dimension of TICBI, see Chap. 6, which discusses the role of culture in planning and intervention.

Other important issues in the pre-planning phase of TICBI assessment involve demographic identification. When working with members of marginalized communities, it is essential to pay attention to their struggles with structural factors that contribute to their active trauma response. For example, evaluating structural inequities as a cause of the precipitating factors through the lens of the intersectionality of oppression is vital for providing TICBI. As Maurice Moreau (1979, p. 78) described it, assessing structural factors for trauma intervention is an approach that allows for the exploration of the intersectionality of oppression, including understanding how specific systems function as trauma triggers, creating stuck points for survivors. It is crucial to remember that structural inequities discussed by Pat Ogden

(2021) create stuck points or delay the recovery process for survivors, continually pushing them beyond their window of tolerance, as shown in Fig. 7.3.

The final step in assessment planning is to explore the cultural factors contributing to your client's trauma response or healing and recovery (see Chap. 6). This includes understanding the specific cultural elements that may contribute to the problem, such as gender-based violence or taboo subjects. Likewise, it is important to plan how to navigate cultural challenges, including discussing cultural taboos during the assessment and intervention, which is influenced by the layered information sharing often utilized by some cultures. This layered sharing (see Chap. 6) favors the in-group (family) as a safe space and protects members from others perceived as an out-group, including the social worker.

To address these challenges, pre-planning to incorporate icebreaker activities in the initial stages of the assessment is crucial. These activities function as the starting point for co-regulation and establishing connections. When planning TICBI, icebreaker activities might consist of a simple information exchange, beginning with discussions about aspects of your client's culture. I often use the icebreakers shown in Table 7.1 to brainstorm and discuss cultural factors without stigmatizing, stressing, or shaming the survivor. Sharing the negative and positive aspects of culture helps to identify trauma responses and risk factors.

Table 7.1 provides discussion points to help you identify each survivor's uniqueness, how they function within the context of biological, psychological, social, and environmental factors, and their service needs during the change talk.

2 Treatment Planning

After completing the trauma assessment process planning, the next step involves pre-planning a tailored intervention. This process includes considering specific treatment issues based on the assessment and exploring applicable trauma treatment approaches (see Chap. 10). TICBI is about improving survivors' quality of life, and

Table 7.1 Identifying survivors' uniqueness

Identified cultural factors	Identified positive culture	Identified negative culture
Micro-level: explore the positive and negative aspects of culture (family, friends, and significant others)		
Macro-level: explore the character-building aspect of your client's culture, influencing their educational aspirations, career choice, the strength of relationships (including tribal), and religious expectations		
Mezzo-level: explore the role of your client's culture in overcoming the systemic and structural of oppression and support systems or available resources		

pre-planning the change talk helps focus on stressful and emotionally overwhelming survival narratives. Therefore, at this phase, plan to further

- Examine the presenting issues and your client's reasons for seeking help to instigate the needed change. Pose questions to gauge their awareness of the pressing problems in their lives
- Explore how adaptive survival responses have become a reality for your clients, who engage in behaviors driven by automatic reactions to avoid danger or escape a situation, thus increasing their vulnerability to mental and physical health issues

Treatment planning involves further exploration of the reasons for seeking services, asking reflective questions, and sharing information while respecting boundaries. TICBI treatment plans emphasize developing rapport and building relationships, which are especially important during the initial stages of intervention. This includes preparing to be the client's ally to motivate, advocate for, and support them as needed, facilitating the transition from PTSD to post-traumatic growth (see Chaps. 3 and 4). Therefore, prepare for motivational interviewing while ensuring the conversation is safe and comfortable to enhance your client's motivation to change. Throughout the treatment, plan to use a language of hope during this process to reduce resistance and guide them toward recovery.

2.1 The Use of Motivational Interviewing for TICBI

In trauma-informed care, motivational interviewing (MI) is a non-confrontational questioning method that can be adjusted for culturally informed communication, fostering hope and reflective conversation. The core motive for using MI is to promote change by recognizing a person's intrinsic motivation to change. It encourages survivors to view events not as the source of their pain or wounds but as part of life, engaging in activities that promote recovery and healing. It motivates clients to move forward and alter maladaptive survival behaviors that contribute to setbacks and impact their well-being.

Motivational interviewing involves using directive questioning to elicit a client's intrinsic motivation for change, which is significant in trauma-informed care (Shulman, 2016). In this process, MI is used to identify survivors' motivations, capacities, and opportunities for change. By employing culturally informed communication throughout the process, MI facilitates an exchange of information based on understanding and compassion, organizes the treatment process, and strengthens the relationship between the client and the social worker.

Pre-planning to use motivational interviewing for TICBI focuses on clearly communicating the issue and the necessary change. MI employs a language of change that helps guide the client away from maladaptive survival narratives and stuck points (chaotic life, self-blame, and anger) toward a focus on change. Pre-planning the treatment reinforces Rogers (1942) reference to unconditional positive regard, as specific conversations about the past and present can be challenging. During these moments, MI facilitates empathy and compassion in the change talk while remaining centered on the issue (Cain, 2010). Treatment planning for TICBI also addresses clients' uncertainty throughout the change process by employing attuned listening and demonstrating empathic understanding of their fear of change. Therefore, using MI

- Encourages clients to engage in reflective discussions about their past and present
- Encourages clients to engage in culturally based change talks, enabling them to view themselves as the experts of their lives and possess the power and capacity to change
- Encourages ongoing change talk during setbacks and helps celebrate comebacks, as the client's mindset fluctuates based on their situation and recent triggers or crises
- Focus on the change plan that matters most to your client now
- Concentrate on your client's values and dedication to change

Since MI involves enhancing motivation and addressing ambivalence, showing interest and curiosity in your clients' stories is important to understand their perspectives, hopes, and dreams for positive change. Planning to use motivational interviewing in TICBI helps you grasp your client's mindset by using reflective and thought-provoking questions, while considering the barriers that taboo subjects create.

In addition, using MI to motivate ambivalent clients, especially those with a survival mindset, requires consideration of potential interaction-related challenges such as transference and countertransference. When working with survivors, there is a risk of triggering memories for both you as a social worker and your client as a survivor of traumatic experiences. Consequently, it is critical to plan how to overcome these challenges. In this pre-planning process, consider how you will respond if your client is dealing with memories, including a fear of authoritative figures that might be triggered (transference). Some clients' reactions can be so intense that they may drive them away from the program. Similarly, think about how you will react if your client's responses or actions trigger implicit memories from your past (countertransference). Either way, it is essential to establish a process for checking these emotions (both yours and your client's) when they arise to avoid adverse effects and treatment breakdowns and to help each survivor based on their trauma responses or mindsets.

2.2 *The Three Survivor Mindsets*

Based on my experiences with the three stages of migration (see Chaps. 1 and 2), I understand the innate human capacity to survive and the motivation to heal, regardless of past experiences of cruelty and loss. However, not all survivors have the same reaction or motivation to heal. Survivors respond to their experiences, and their brains and bodies organize or prioritize mobilizing their defense system or survival responses in anticipation of danger (Ogden & Fisher, 2015; van der Kolk, 2014).

I began to categorize survival responses into three mindsets after paying attention to clients' responses and observing my children's reactions to their traumatic pasts. These observations motivated me to conduct a study during the height of the 2020 COVID-19 pandemic entitled "Re-Thinking Resettlement and Integration: Creating Trauma-informed Culturally Based Intervention or (TICBI)." To explore survivors' mindsets during their forced migration and resettlement (see Chaps. 1 and 2), I recruited 20 survivors and 23 service providers working with survivors for individual interviews and focus groups. In this study, I learned from practitioners about the various culturally responsive approaches they use to improve the one-size-fits-all service model and how they create inclusive approaches to fit the diversity of the population they serve.

Understanding survivors' mindsets is vital for treatment planning because emotional reactions and sensations from past trauma activate the vagal pathways, creating an autonomic nervous system (ANS) survival response. Planning TICBI involves recognizing these involuntary responses and their adaptive sources, often shaped by past experiences, culture, personal values, or religious teachings. Based on multiple supportive or triggering factors, survivors develop emotional responses and body/brain reactions that shape their mindset. In this chapter, I categorize these mindsets into three survivor types based on my observations and the descriptions provided by the research participants from the survivors' group as they share their narratives of survival, struggles, and healing processes.

2.2.1 Survival Mindset

As the first stage of trauma response (fight, flight, or freeze), survivors in this mindset focus on their past and present traumatic pain. They may become stuck using maladaptive responses, which continue to create challenges that affect their well-being. Survivors with this mindset live in a constant state of alertness and operate with brain/body emergency stress responses (sympathetic mobilization). Survival is a temporary human reaction to life-threatening situations. However, it creates chronic hypervigilance or reactivity when it becomes an emotional impasse, impacting survivors' day-to-day lives. Therefore, pre-planning the treatment process helps address this group's emotional responses, including over-vigilant behavior and an intense survival mindset and reaction (fight, flight, or freeze [dorsal vagal collapse]) that affect their well-being.

2.2.2 Therapeutic Mindset

This is the second stage of trauma response, where the survivor develops a post-traumatic positive attitude that reduces their traumatic reactions and encourages them to form social connections. Although not fully recovered, participants in my study discussed how they used their cultural and religious teachings to foster self-compassion, a strong faith, and an understanding of their triggers. This helped them manage their emotions more effectively and increased their motivation to seek help. Religion inspires them to practice self-compassion and reflection, enabling them to interpret events based on facts rooted in education, culture, and religion. This process helps them make sense of their experiences, find meaning, and identify a purpose for healing through connection. Being in a therapeutic mindset allows them to regulate their emotions, promote neuroplasticity, and create new neural connections, supporting healing and offering opportunities for growth and transitioning to an active, engaged mindset.

2.2.3 Active, Engaged Mindset

Once they transition to this mindset, survivors are well on their way to recovery. Thus, when pre-planning treatment for this group, it is crucial to remember that, regardless of the cruelty, violence, loss, and displacement of war, this group of survivors has overcome those challenges and is using their natural strength and capacity to thrive in post-traumatic actions. The participants in my study repeatedly commented on how they reach the stage of recovery by utilizing their internal strength and resilience, and how cultural and religious learning enables them to envision a better tomorrow. For most survivors, it is an internal acceptance motivated by cultural values and environmental resources, which contributes to managing intense feelings associated with the traumatic past. The combination of these factors helps survivors control their overactive responses by activating their ventral vagal regulation and deactivating their sympathetic nervous system response (fight, flight, or freeze) to promote a sense of connection. When participants were asked about the beginning of this transformative acceptance, they described various supportive systems, but an important one was their internal resilience. These motivating and supportive factors enabled them to develop a new neuronal assembly, repairing their disconnected neurons (also known as pruned or underused neurons in the amygdala and hippocampus) and, in some cases, breaking through chronic trauma that had kept them frozen in a survival mindset. The following figure shows emotional reactions related to the three mindsets that are the basis for understanding trauma response (Fig. 7.1).

2.3 *Planning Treatment*

When planning TICBI for clients with one of the three mindsets, it is fundamental to pay attention to their responses and positions on the mindset scale. The availability or lack of services can create stress that causes them to fluctuate on this scale. Figure 7.2 demonstrates how survivors' mindset sometimes shifts due to unexpected stressors that disrupt their support systems and sources of emotional regulation,

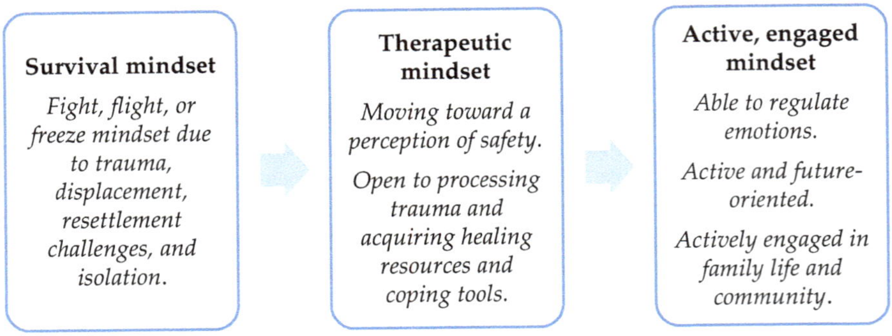

Fig. 7.1 The three survivor mindsets (Bokore et al., 2023) (Adapted from Bokore et al. 2023, Figure 2. Minor modification was made. https://doi.org/10.5296/ijsw.v10i1.20561, licensed under the terms of the Creative Commons Attribution 3.0 Unported License (https://creativecommons.org/licenses/by/3.0/).)

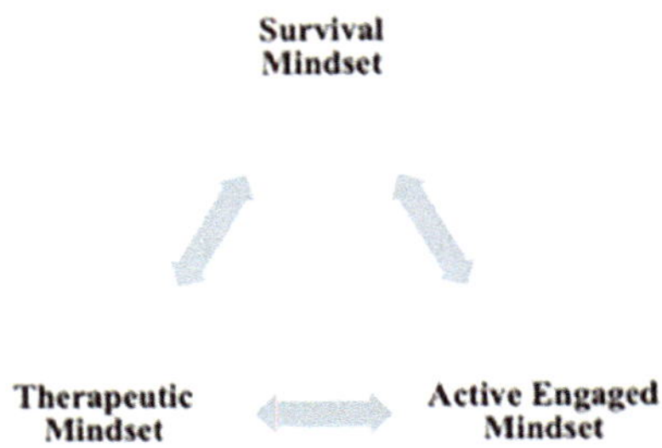

Fig. 7.2 The cycle of survivor mindsets (Bokore, 2022) (Developed by the TICBI Project, this illustrates the impacts of stressors causing survivors to move between the cycle of mindsets. The figure was part of the presentation by the author at the 19th Online Conference (August 1–5, 2022) of the International Association for the Study of Forced Migration (IASFM), hosted by the Catholic University of Santos, Brazil.)

altering the course of autonomic responses and keeping them trapped in the cycle of mindsets.

Planning to engage survivors in evaluating their active, engaged mindset requires understanding each survivor's mindset based on autonomic responses. As a conversation starter, I used a collaborative drawing, such as a genogram or culturagram, to determine where survivors based their emotions, available support systems, and triggers. This collaborative drawing provides a window into their mindset, revealing their stressors and the extent of their emotional regulation. Plan to use collaborative, culturally informed motivational interviewing to ease into an exploratory conversation about the mindsets and healing process. As you prepare for treatment planning, think about how you will

- Explain the intervention process, detailing its goals and steps to achieve the desired change

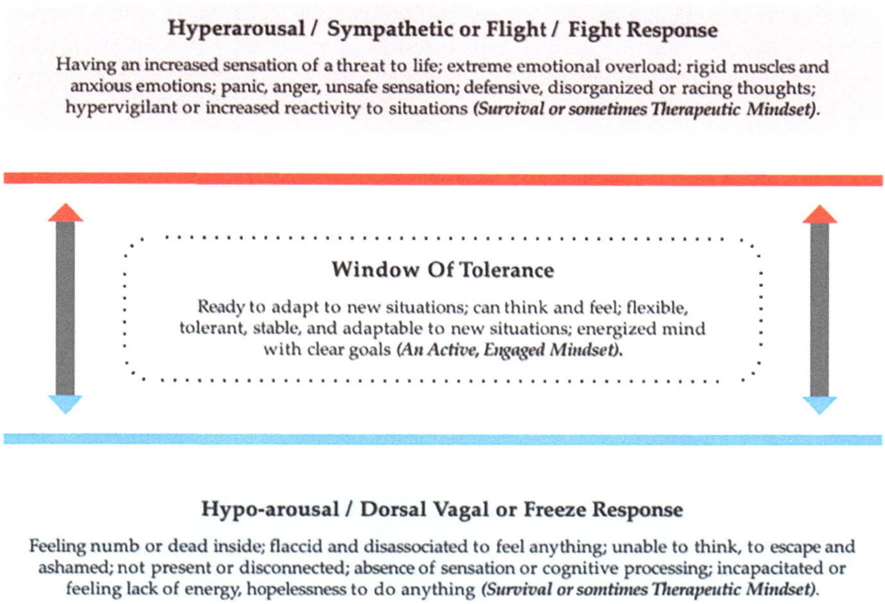

Fig. 7.3 The window of tolerance (Bokore, 2022) (The figure was part of the author's presentation at the 19th online Conference (August 1–5, 2022) of the International Association for the Study of Forced Migration (IASFM), hosted by the Catholic University of Santos, Brazil.)

- Create an environment where cultural awareness is prioritized, and personal, physical, or emotional issues are discussed through respectful information exchange
- Be mindful of your emotions as you prepare for the challenging yet rewarding process of providing TICBI based on the survivor's mindset. Remember the theories that will guide the intervention, enable you to follow the person-centered approach, and view the issues from the "person in the environment" Rogerian perspective

2.3.1 Treatment Planning for Clients with a Survival Mindset

In trauma care, paying attention to autonomic responses is a good starting point. Trauma reactions are implicit, and Stephen Porges first described these survival-based reactions led by the autonomic system in the 1990s. When I first read Porges's (2007) work, my response as a survivor of multiple traumas was self-understanding, which came with self-compassion as I now understood my reactions to specific triggers. Porges's polyvagal theory delineates human survival actions and reactions. For some survivors, the reaction becomes their stuck point, led by the oldest part of the brain, continuously reacting to stimuli or triggers (sympathetic response—fight, flight, or freeze). Planning an intervention for someone functioning with a survival mindset requires patience, as you will listen to your client's survival stories told multiple times (they are recounting stuck points and seeing danger everywhere) or encounter your client's overreactive or exaggerated negative thoughts about themselves or others. Survivors often share their thoughts through a narrative of survival with repeated descriptions of physical responses (e.g., hyperarousal and sympathetic response, increased heart rate, fibromyalgia, headaches, and digestive problems) or psychological responses (experiencing panic attacks and having difficulty focusing and retaining information). In the preparatory planning phase, remember the possibility that structural or systemic issues can trigger your client. Survivors from marginalized groups often deal with identity issues or barriers to group membership in a post-colonial country, increasing the "traumatogenic effects of the intersectionality of oppression or insidious trauma" (Watson et al., 2016, p. 656).

In the context of war, oppressed groups remain stuck in a survival mindset. They function from the old, but fast, primitive brain (limbic system and brainstem response) that they use to overcome the threat they face. However, the chronic use of this natural, lifesaving tendency (fighting, fleeing, or freezing in the face of danger) creates stuck points for those with the survival mindset, affecting their quality of life. With this in mind

- Create a process or activity to evaluate your client's responses and identify significant physical changes in their physical tension (e.g., sweating or feeling cold) when they respond to your questions
- Be attuned to survivor stories and listen for behavior descriptions or actions that indicate your client's fight, flight, or freeze response (hyperventilation, emotional numbness, and hopelessness)
- Be attentive to their repeated survival stories and explore their meaning-making processes
- Listen to how they tell stories of cultural shocks and the expression of anxiety, stress, helplessness, irritability, and how much they miss their old life
- Pay attention to stories showing stuck points or descriptions of physical reactions based on frequent sympathetic responses (e.g., rising heart rate, which delivers blood and oxygen to the body in preparation for escape)

Treatment planning begins by acknowledging the innate human biological need for connection and, as Chap. 2 demonstrates, understanding your client's journey through war, forced migration, border crossing, and resettlement. Survivors with a survival mindset find moving away from protective, sympathetic responses frightening and challenging. Therefore, consider your client's struggles and work toward change. This requires a positive therapeutic relationship to slowly move them toward change and activate their parasympathetic and ventral vagal response. When planning an intervention for survivors with this mindset, pay attention to your client's cultural background; trauma response and cultural variation make a difference in the meaning-making and recovery process (see Chap. 6).

Planning helps you recognize cultural variations, such as individualistic versus communal cultural responses. For example, while a person from an individualistic culture sees each human as a "single, isolated organism," someone from a communal culture recognizes human interdependence, relying on emotional, mental, and physical well-being (Cozolino, 2010, p. 178). This perspective makes a difference for your client in imagining the availability of resources and support systems. To explain the importance of social connectivity in the healing process, you can use the example Cozolino (2010, p. 178) uses of children in orphanages who are deprived of human connection to show how isolation kills and how socialization, or human connection, saves lives.

An effective intervention involves recognizing these differences and needs by acknowledging your client's stories. It includes showing empathic understanding and being attuned to the present as your client shares their emotions, traumatic experiences, and body/brain responses. Having a plan helps reduce stress and anxiety related to meeting your client for the first time. It also helps in co-regulation and providing your client with the necessary psychoeducation to understand their situation (see Chaps. 3 and 4). To prepare for the upcoming TICBI, it is also essential to identify the resources you need to work with your client. This can include internal and external consultation with experts or experienced social workers.

For clients with a survival mindset, day-to-day survival functioning is stressful and often isolating, especially for those dealing with PTSD (Chap. 3). Remember to be patient as it takes time to acquire therapeutic alliances, foster open and culturally attuned communication that encourages collaboration, and establish realistic expectations about the intervention and healing process. Cultural communication makes a difference in intervention planning (see Chap. 6). For example, if your client is Buddhist, plan to use encouraging words such as *metta* (loving-kindness). Buddhist participants in my TICBI study talked about how they use this practice for balance, self-kindness, and calmness. They said the process helps them have a nurturing and peaceful mind, showing compassion to themselves and others. Similarly, the Muslim participants stated that what helps them achieve a calm mind is reciting verses from the Qu'ran, rolling their prayer beads in their fingers, or participating in the religious ritual dhikr, a Sufi prayer for spiritual perfection. The Christian participants described their religious rituals, practices, and belief systems throughout their displacement journey.

Understanding your client's cultural values and traditional practices will help you avoid using emotionally triggering words that may impact the therapeutic relationship. Awareness of your client's cultural values and primary need to feel safe is essential in providing TICBI. In your intervention or treatment planning, the aim is to move your client from a survival mindset to an active, engaged mindset using culturally acceptable conversations that help them manage their emotions and reactions.

2.3.2 Treatment Planning for the Therapeutic Mindset

Intervention planning for those with a therapeutic mindset begins with recognizing your client's reactions to past trauma and their rationale for the issues they are dealing with. While working with survivors with this mindset, identify their internal working models, or mental images of themselves, which are helping them move toward safety and connection. The development of positive mental representations begins at an early age, but it continues to change throughout life. For those with this mindset, these representations help them see hope and survive in challenging times.

Survivors with this mindset have a flexible outlook on life, indicating their readiness to embrace possibilities, whether accessing available information or establishing social connections. They use self-compassion to understand the past, accept the present, and move toward change. Intervention planning for those with this mindset begins by

- Focusing on precipitating/triggering factors, including recent stressors and triggers
- Planning an in-depth exploration of the perpetuating/maintaining factors contributing to the cycle of changing mindsets (from a therapeutic mindset back to a survival mindset), decreasing your client's motivation to change

Planning an intervention for clients with a therapeutic mindset includes thinking ahead about activities that can be used to explore their values, what is important to them at this moment, and what motivates them to work on their healing. Plan questions you might ask and how you will address specific issues as you explore your client's feelings, emotions, attitudes, and self-regulation, which can lead to adaptation or motivation to change. In this plan, awareness of your client's traumatic responses, culture, and motivation for healing is key. Understanding how to use motivational interviewing skills in this process, including discussing cultural and religious values, can help your client reframe their thoughts to focus on what is important to them as they continue to work on change (Gordon & Borushok, 2017). Planning to use a client-centered approach to understand their worldview and what is essential to them moves the conversation to a forward-looking exchange (Gordon & Borushok, 2017; McLean & Follette, 2015; Miller & Rollnick, 2002).

In planning an intervention for those with a therapeutic mindset, it is essential to remember that trauma is a wound (Chap. 3) that reopens from time to time when clients encounter triggers created by barriers and stressors, including lack of resources or support systems. Planning an intervention informed by solution-focused therapy is a goal-oriented approach that identifies survivors' past capacity to overcome challenges and motivates them to change. Clients with a therapeutic mindset use nurturing techniques they learned from their parents, culture, and extended family to help them develop secure attachments and internal resilience (Schore, 2012). It is also essential during the planning stage of the intervention to consider the impact of stress, which at times becomes unbearable even for those with a therapeutic mindset. When working with marginalized survivors, plan to use stories of hope based on your client's values, motivation, and goals.

Using culturally based metaphors to describe your client's setbacks and healing helps lighten the conversation and create co-regulation to discuss further values, motivation, and action (Gordon & Borushok, 2017). Culturally based metaphors help in collaboratively examining what it would be like to move beyond the cycle of moving forward and backward based on triggers and living on an emotional island circling from hyperarousal to hypoarousal and away from the optimal emotional zone, that is, their "window of tolerance" (Ogden, 2021). Therefore, when planning an intervention for those with a therapeutic mindset

- Pay attention to the precipitating factors or underlying reasons for your client's emotional swings (between survival and therapeutic mindsets) and how they affect their autonomic responses
- Plan to include mindfulness practices, focusing your client's attention on the present moment without judgment or blame, regulating their emotions, and finding solutions for their current issue causing the setback

Confident in the human capacity for survival, plan to include conversations based on hope, values, and motivation, using the language of religious or cultural acceptance (e.g., God's or Allah's acceptance[1] or tribal pride). Focusing on factors in your client's life that show internal resilience helps them move back to recovery and a therapeutic mindset, and even shift to an active, engaged mindset.

In this planning phase, include culturally informed regulatory exercises (see Chap. 10) to reduce stress or provide psychoeducation to help your client understand their situation. Engage in empathic and reflective conversation and give constructive feedback to move your client toward an active, engaged mindset.

2.3.3 Treatment Planning for the Active, Engaged Mindset

Similar to those in the therapeutic mindset, intervention planning for clients with an active, engaged mindset focuses on stressors that create setbacks (precipitating or triggering factors), interrupting your client's post-traumatic growth of feeling secure and having access to resources (see Chap. 4). Therefore, assessment planning for those with this mindset includes exploring their cognitive processes and issues that increase their vulnerability and setbacks. This vulnerability may be tied to their neuroception, where their autonomic nervous system evaluates what is happening in their internal organs (stress responses) in response to environmental stressors. This is an ongoing evaluation of responses to implicit body/brain reactions, including danger alerts. Therefore, planning an intervention for such clients requires an understanding of their shift from functioning in the window of tolerance to either a dorsal vagal state of immobilization and collapse or a sympathetic response (fight and flight) moving away from their feeling of safety and active ventral vagal system of being ready for social engagement and managing their lives.

Intervention starts with an attuned and compassionate conversation about pressing issues, using culturally informed communication about the past and present and what brought your client to seek services. It includes planning motivational

[1] Those who believe in their faith or religion during life's challenges often turn to their God, Allah, or any other deity they worship.

interviewing to assess your client's cognitive processes and behavioral responses to continue with the change talk.

- Explore your client's values, motivations, and commitment to healing and rebuilding their lives.
- Explore your client's goals and values that actively helped them seek support, growth, socialization, and integration, including their adaptive responses to the behaviors contributing to their coping mechanism.
- Explore your client's past and present using culturally informed motivational interviewing to discuss future goals.
- Listen to your client's future-oriented stories, note their motivation when dealing with their personal life, family, and community commitments, and ask how they overcame specific barriers. Past strengths will inform future change, assist the intervention process, and help plan the change talk.

Even though survivors with an active, engaged mindset are familiar with establishing a social connection and seeking support (see Chap. 4), they may also be dealing with precipitating factors and need you to listen to their stories, revealing both their emotional dysregulation and vagal response. Plan to establish connections, identify helpful resources, and share information. For those with an active, engaged mindset, rapport-building and establishing community connections for resources and additional support are necessary.

Planning an intervention for those with an active, engaged mindset includes assessing their recent challenges in social relationships. Like you did for the other two mindsets, plan to establish a solid collaboration and connection between you and your client. Pay attention to the stories of new stressors that reverted them from an active, engaged mindset to a therapeutic or survival mindset. If some of those stressors are related to social, economic, or structural factors, be prepared to take on an advocacy role to support them with access to a community, service agencies, and resources. Incorporate other activities, such as group work, that will strengthen their resiliency, re-establish their confidence, and allow them to continue with their survival actions, contributing to actions of post-traumatic growth.

3 Pay Attention to Variation

When working with clients, regardless of their mindset, it is critical to provide services that do not impose foreign values and not to offer suggestions that include quick adaptation to the new culture. A Syrian refugee research participant in my TICBI project expressed their frustration on this point: "Resettlement based on cultural orientations seems to suggest I need to quickly adapt to whatever they were telling me to assimilate into the right way of living." Adding information sharing (psychoeducation) into your interventions should be culturally congruent with your client's needs. While planning to work with survivors, always remember that a respectful discussion where their values and priorities are at the center of the conversation moves your client forward and strengthens the social worker–client relationship. Working with someone in an active, engaged mindset requires awareness of their traumatic experiences and the culturally adaptive survival responses and social-emotional skills that enable them to function within their window of tolerance.

4 Pay Attention to Brain Responses for Each Mindset

Survivors of prolonged wars have wounded souls and are vulnerable to anxiety. Trauma responses leave survivors with a low threshold for stress, creating extreme hyperarousal or hypoarousal (Levine, 2015, p. 43) or narrowing their window of tolerance (Ogden & Fisher 2015, p. 73). For survivors with the survival and therapeutic mindsets, past experiences make them vulnerable to emotional swings. Planning for TICBI needs consideration of these emotional instabilities, expanding your client's window of tolerance.

Your intervention plan must include a careful evaluation of emotions and an exploration of the reasons behind your client's sensitivity to triggers, narrowing the size of their window of tolerance while expanding their hyperarousal or hypoarousal as demonstrated in Fig. 7.3. Survivors of chronic trauma often show signs of recovery from their past ordeals. However, at times, they may also regress, moving from an active, engaged mindset to a therapeutic mindset or even back to a survival mindset. For these survivors, being in this cycle indicates that they are functioning outside their window of tolerance, disrupting the adaptive responses that previously allowed them to work toward developing post-traumatic growth.

To break the cycle of back-and-forth within the survival mindset and help your client move forward, create activities that re-establish balance in their lives while re-engaging their self-regulation and homeostatic responses needed to keep them at the threshold of their window of tolerance (see Chap. 10).

Re-engaging the self-regulation process and homeostatic responses widens your client's window of tolerance, a process informed by applicable theories that help you assess issues impacting your client's well-being in any of the three survival mindsets. Considering one or more theories that will inform the intervention is essential. Social workers often adopt a generalist approach that includes the flexibility to choose theories to help them explore the issues and plan an intervention based on their client's innate capabilities. This also involves planning for foreseeable barriers, including addressing transference and countertransference issues.

Providing TICBI in a social work setting also requires reflection on the intervention process and consideration of the interaction between you and your client. This means recognizing the self while providing TICBI services. This self-reflective practice will help you identify transference and countertransference as they occur. It will also alert you to feelings evoked by your work or the issues you are exploring, and provide a better understanding of how it stimulates implicit emotions in both your client and yourself.

5 Paying Attention to Transference and Countertransference

5.1 *Transference*

Transference is a client's cognitive and physical reaction to the therapeutic relationship and responses to a word or misinterpretation of an action or reaction. Cultural diversity and traumatic experiences sometimes play a part, leading clients to react, whether consciously or unconsciously, to an expectation they have about the social worker or service. Some of these reactions may involve transferred emotions from their past experiences, including relationships with other social workers or individuals in positions of authority.

Recognizing these emotions or reactions helps social workers prepare to address client perceptions about the work, the relationship, and the outcome. It is important to pay attention to these emotions when working with clients who have experienced migration and the associated trauma, as they may impact or jeopardize the therapeutic relationship. For survivors, the fear of being hurt again by those in authority often takes over, leading them to withdraw from a program and, in some instances, generate feelings of countertransference.

5.2 *Countertransference*

Providing TICBI services involves examining your client's adaptive responses and behaviors that damage or break their relationships with others and reflecting on your inner systems based on personal experience. Countertransference, the emotional reaction of the social worker and the redirection of those emotions to the client, requires attention during this planning phase. Avoiding countertransference begins with acknowledging your reaction to your client's feelings of depression or symptoms of hypoarousal and the impact these have on them. Feelings that served as adaptive means of survival during forced migration may have been transferred to a fear of the failure of the therapeutic relationship.

Similarly, to avoid misinterpretation and reactions to your client's hypoaroused state, numbed feelings, or lack of emotional connection with you, it is essential to remember your client's experiences during the first two stages of their forced migration (see Chap. 2), such as anxiety, numbness, grief, and loss, which are transferred to their current emotions. This may also include displaying angry outbursts or hyperarousal, which they use as a protective measure to push people away before they can cause harm or, worse, commit physical or sexual violence against them or their family members.

To avoid adverse effects, create checks and balances to assess the presence of transference and countertransference. During the planning stage and intervention, be alert for emotional conversations that may create emotional entanglements that affect the therapeutic relationship. Plan the intervention to include a mechanism for checking fear-based responses and engaging in empathic conversation with attuned listening, which helps avoid entanglements and service breakdowns. It is also essential to identify cultural barriers that create problems for your client because service breakdown is not always due to transference:

- Some barriers are misunderstandings created by your client's cultural expectation that leaders and authority figures, in this case, you, are knowledge holders.
- Encourages knowledge sharing and collaborative work, limiting advice given only when asked to strengthen the relationship, build trust, and avoid creating a relational breakdown, which is often mistaken as an issue of transference.

Planning helps decrease the possibility of transference and countertransference. It motivates social workers to engage in self-reflective activities, including attending training or supervisory meetings, to discuss these issues and remedies or preventative actions.

6 Pre-planning for Theories That Can Inform Your Intervention

Another critical intervention process in TICBI planning is deciding what theories or perspectives will inform the intervention.

6.1 Strengths-Based Perspective

One of the theories often considered in social work practice is the strengths-based approach, which originated in Jane Addams's work in a resettlement house in the United States in the early twentieth century. Using this approach in planning the intervention guides social workers in building on the strengths, capabilities, and skills retained by survivors while focusing on change talk that promotes well-being. This approach allows social workers to explore their clients' resources, such as educational background and social support, to promote well-being. The perspective helps avoid focusing on deficits instead of your client's positive attributes that will contribute to change.

A strengths-based approach is about paying attention to people's positive qualities, especially when working with marginalized communities often labeled as having deficits. In these instances, the approach helps to evaluate the intersectionality of oppression that contributes to their ongoing trauma. The approach acknowledges external barriers and internal strengths, emphasizing the role of resilience, social networks, skills, and talents rather than deficits to help survivors engage in change talk and move toward growth and healing.

6.2 Critical Ecological Systems

It is human nature to strive to improve and adapt to one's environment. Despite their horrific past, survivors are motivated to be optimistic about their capacities when they see opportunities and resources within their environment. Reciprocal engagement with their environment leads to a therapeutic or active, engaged mindset. The environmental fit makes them use their essential human capacity and motivation. Unfortunately, when environmental and personal limitations are triggered by oppressive systemic structures and social and physical environments, their heroic survival and adaptation change to stressful, maladaptive, and impaired functioning. When these stressors trigger survivors who are otherwise doing well to sense the imbalance between their environment and their basic needs (shelter, safety, employment), they move back emotionally to their survival response, which only adds to the barriers they are experiencing.

The first social work course recognizing the connection between individuals and families and their environments (an early form of critical ecological systems) was taught at Columbia University School of Social Work over 100 years ago (Wood & Geismar, 1989). The critical ecological systems perspective builds on the strength-based perspective (see Chap. 10) and a belief in human capacity. It allows social workers to explore survivors' positive reciprocity with their environments. Using critical ecological systems in the TICBI intervention plan helps provide a holistic framework to view the person-in-environment by exploring survivors' issues, adaptability, and resiliency.

This perspective allows service providers and clients to collaboratively examine the impact of structural and systemic societal trauma, such as racism, sexism, and poverty. Instead of focusing on the survivor's deficits, critical ecological systems provide a framework for looking at their problems from a societal context, such as the systemic and structural issues that overwhelm and trigger your client's trauma. Thus, incorporating critical ecological systems into your intervention planning offers a valuable resource for understanding the intersectional nature of trauma for marginalized populations, particularly for those with a survival mindset, and sometimes for those with a therapeutic mindset, who are stuck in a cycle of survival narratives. I use the word *sometimes* for the latter group because they are often able to draw on their internally resilient behaviors, relapsing only into a survival mindset because of new triggers or stressors.

6.3 The Narrative Perspective

In TICBI, the narrative approach (discussed in Chap. 10) allows social workers to explore the intersectional nature of trauma, encouraging them to engage clients in a respectful, non-pathologizing, and non-blaming culturally informed conversation where their concerns and voices can be heard. Even though the approach has its challenges when working with survivors with intellectual disabilities or with those with language barriers, when appropriate, it helps to explore in a safe space past experiences and identify barriers to and strengths for change. It allows survivors to be the authors of their stories, moving them from problem-saturated stories to healing narratives that remove them from a cycle of problematic narratives.

Incorporating the narrative approach into TICBI also helps to focus on your client's stories of protection (sympathetic or dorsal vagal collapse) and connection (active ventral vagal response) to understand their interpretation of events and actions (Larsson & Sjöblom, 2010; Riessman, 2005; Siegel, 2011). The narrative approach goes beyond interpretation and actions, bringing out your client's voice and views about the war/conflict they lived through and their displacement. It also allows you to explore your client's stories while considering the context that shapes them. It encourages attuned listening to identify problem-saturated stories (stuck points) in your client's narratives, impacting their move toward growth, establishing connections, and engaging in the healing process.

6.4 Acceptance and Commitment Therapy

Acceptance and commitment therapy (ACT) in TICBI helps to focus the conversation on the present moment and survivors' narratives to evaluate their meaning-making processes and make a change (Hayes, 2004, 2016). In TICBI, the use of acceptance and commitment therapy benefits particularly survivors who believe in a higher power and religious teaching and those conditioned to a survival culture (marginalized groups). For this group, acceptance and commitment therapy allows them to focus on teachings from their culture or religion on what matters most to them. It is a value-based, motivational intervention rooted in acceptance, strength, and survival as the guiding principles of life, structured around spiritual and cultural values. Acceptance and commitment therapy helps survivors focus on what kind of family or religious or community member they want to be, what is deep in their hearts, and what matters the most to them. Based on cultural or religious teachings, what matters the most may differ; however, increasing your client's motivation depends on their ability to change and move toward helpful actions in their healing process and to rebuild their lives. In TICBI, acceptance and commitment therapy is used as a collaborative conversation to include your client's strengths, work on meaning-making, and increase the possibility of making a meaningful connection that is helpful for their recovery.

Acceptance and commitment therapy is a tool for social workers to refocus on social work's core value of being attuned to survivors' needs in providing services, hearing their views and what matters most to them with openness, compassion, and curiosity. When using acceptance and commitment therapy, social workers use one of the six acceptance and commitment therapy processes called *cognitive diffusion*. This is used for looking at memories while interrupting the negative thoughts attached to trauma wounds and narratives to minimize their impact and work toward making a change (Hayes, 2004, 2016). Similarly, including *acceptance* is another part of acceptance and commitment therapy. It motivates survivors to explore their inner thoughts, feelings, sensations, memories, and reactions to help them manage their day-to-day lives (Gordon & Borushok, 2017, pp. 40–42). Using these processes in TICBI allows survivors to practice flexibility of thought (increased self-reflection) and choose to respond positively to situations, redirecting their emotions and vagal responses simultaneously (from sympathetic, poor outcomes to parasympathetic, positive outcomes).

6.5 Polyvagal Theory

In the past two decades, clinicians have turned to the vagal path to define trauma response, including the need for connection, protection, and disconnection based on an autonomic response to the environment (Cozolino, 2010; Levine, 2015; Porges & Dana, 2018; Siegel, 2011; Sweeton, 2021). When planning TICBI, understanding the functions of the vagal path, or the "vagal circuit" (Porges, 2007), helps explain to survivors their body/brain reaction, moving away from self-blame to self-understanding. The central organizing principles of polyvagal theory define these survival systems and structures, which regulate involuntary physical responses and the functions of the sympathetic and parasympathetic nervous systems, to preserve life and maintain homeostasis (Dana, 2020).

As mentioned in Chap. 3, the autonomic nervous system regulates involuntary physiological responses (e.g., heart rate, blood pressure). It creates a sympathetic response, a way to avoid the anticipated danger, and energy to seek the opportunity to escape. When the sensation of safety is felt, the parasympathetic system is activated, promoting social engagement (Porges, 2007). These natural survival activities are the source of the built-in surveillance system (neuroception) that persistently interprets incoming information through physiological safety sensors. In providing TCIBI, explaining these mechanisms is part of creating collaboration, connection, and co-regulation, helping survivors decrease danger alerts, work on recovery, and broaden their window of tolerance. In providing TICBI, understanding and explaining these mechanisms are part of collaborating, creating connections, and co-regulation, helping survivors decrease danger alerts, work on recovery, and widen their window of tolerance.

Planning TICBI requires a clear understanding of how the human surveillance system (neuroception) functions to keep us safe and how it can be used for self-understanding and motivation to work toward healing. Understanding the vagal path in trauma response and working toward healing includes introducing mindfulness practices that help to create a balance (homeostasis) or widen the window of tolerance, which is necessary for promoting a functional move toward an active, engaged mindset.

7 Mindfulness

In planning TICBI, incorporating mindfulness (see Chap. 10 for more details) into the intervention is necessary for emotional balance. Using culturally acceptable communication for instruction in meditative practices increases emotional balance and healing. In TICBI, mindfulness practices are used to help the client move away from judgment (self-protection or sympathetic mobilization) and adverse trauma (self-blame, guilt) to paying non-judgmental attention to the past (acceptance) and focusing on the present to promote growth, change, and healing (Gordon & Borushok, 2017; Kabat-Zinn, 2005; Siegel, 2021).

Definitions

Acute stress disorder (ASD) is a PTSD-like trauma response that lasts for a short period, typically a few days to weeks.

Adjustment disorder: Experienced only during periods of stress or life changes, characterized by anxiety and worry regarding the issues or changes involved.

Change talk is a careful conversation structured as a motivational interview that thoroughly examines problem-saturated stories, situations, and contexts. It focuses on the survivor's interests, motivations, and preferences, utilizing motivational language to understand the issues, reduce resistance, and foster connection and growth/change.

Co-regulation occurs when the nervous systems of two individuals interact (through mirror neurons) to support one another and emotionally connect through mutual regulation of their physiological states.

Culturagram: An assessment tool that offers a graphical representation of various aspects of survivors' lives, including family structures, relationships, and the rules utilized for assessment and intervention planning.

Genogram: A visual method for gathering information to better understand survivors' relationship dynamics and behavioral patterns, promoting self-awareness and assessing support systems and problem areas to guide the intervention process.

Motivational interviewing is a component of change talk. It facilitates a collaborative conversation by engaging survivors in the change process and focusing on the presenting issues. Motivational interviewing stimulates creative thinking, reflects on past survival skills, and aids in planning for change.

Transference and countertransference: Transference refers to the client's redirected emotions and feelings toward the social worker, while countertransference involves the social worker projecting personal feelings or unresolved conflicts onto the client. Both reactions affect the quality of the working relationship.

Traumatogenic: "Relating to the dynamics by which a traumatic event (e.g., childhood sexual abuse) may have long-term negative consequences, including the development of a mental disorder" (APA Dictionary of Psychology, 2025).

Window of tolerance: The optimal range of emotional arousal for achieving emotional balance and effectively managing daily activities.

Discussion Questions

1. Summarize each section of this chapter. What are the key pre-planning suggestions in each section? Include a quotation from this chapter that resonates with you.
2. What information should you gather for each identified mindset to explore your client's presenting problem or concern that brought them to your practice?
3. What information do you need to identify your client's strengths and resources based on their mindset, and how will you request it? Provide examples using motivational interviewing.
4. What must you know about bio-psychosocial and structural stressors in your client's life?
5. Review the three TICBI mindsets and discuss the primary objective of pre-planning assessments and interventions for each mindset.

Self-Exploration Questions

1. Reflect on a time when you sought services for school or other social programs, and describe the assessment you underwent to receive those services or enter the program. What aspects of the assessment made you feel comfortable or uncomfortable? Why?
2. How will you use your personal experiences to plan an intervention and assist your clients?

Competing Interests The author has no conflicts of interest to declare that are relevant to the content of this chapter.

References

APA Dictionary of Psychology. (2025). *Traumagenic*. https://dictionary.apa.org/traumagenic.

Bogo, M. (2006). *Social work practice: Concepts, processes, and interviewing*. Columbia University Press.

Bokore, N. (August 1-5, 2022 - Online). *Re-think Resettlement and Integration: Creating Culturally Based Trauma-Informed Interventions*. 19th Conference of the International Association for the Study of Forced Migration (IASFM), hosted by the Catholic University of Santos, Brazil.

Bokore, N., McGrath, S. L., McGuire, P., Karod, A., Rahimpour, M., & Asokumar, A. (2023). Developing a trauma-informed culturally-based intervention (TICBI) approach for refugee resettlement practices. *International Journal of Social Work, 10*(1), 1. https://doi.org/10.5296/ijsw.v10i1.20561

Cain, D. J. (2010). *Person-centered psychotherapies*. American Psychological Association.

Cozolino, L. J. (2010). *The neuroscience of psychotherapy: Healing the social brain* (2nd ed.). W. W. Norton & Company.

Dana, D. (2020). *Polyvagal exercises for therapists and clients: 50 Client-Centered Practices*. W. W. Norton & Co.

Edwards, S., Edwards-Stewart, A., Dean, C., & Reddy, M. K. (2023). Evaluation of posttraumatic stress disorder and acute stress disorder VA/DOD clinical practice guidelines training. *Military Medicine, 188*(5–6), 907–913.

Gordon, T., & Borushok, J. (2017). *The ACT approach: A comprehensive guide for acceptance and commitment therapy*. PESI Publishing & Media.

Hayes, S. C. (2004). Acceptance and commitment therapy, relational frame theory, and the third wave of behavioral and cognitive therapies. *Behavior Therapy, 35*(4), 639–665. https://doi.org/10.1016/s0005-7894(04)80013-3

Hayes, S. C. (2016). Acceptance and commitment therapy, relational frame theory, and the third wave of behavioral and cognitive therapies – republished article. *Behavior Therapy, 47*(6), 869–885. https://doi.org/10.1016/j.beth.2016.11.006

Kabat-Zinn, J. (2005). *Coming to our senses: Healing ourselves and the world through mindfulness*. Hyperion.

Larsson, S., & Sjöblom, Y. (2010). Perspectives on narrative methods in social work research. *International Journal of Social Welfare, 19*(3), 272–280. https://doi.org/10.1111/j.1468-2397.2009.00672.x

Levine, P. (2015). *Trauma and memory: Brain and body in a search for the living past: A practical guide for understanding and working with traumatic memory*. North Atlantic Books.

McLean, C., & Follette, V. M. (2015). Acceptance and commitment therapy as a nonpathologizing intervention approach for survivors of trauma. *Journal of Trauma & Dissociation, 17*(2), 138–150. https://doi.org/10.1080/15299732.2016.1103111

Miller, W. R., & Rollnick, S. (2002). *Motivational Interviewing, second edition*: Preparing People for Change. Guilford Press.

Moreau, M. J. (1979). A structural approach to social work practice. *Canadian Journal of Social Work Education, 5*(1), 78–94.

Ogden, P. (2021). *The pocket guide to sensorimotor psychotherapy in context (Norton Series on Interpersonal Neurobiology)*. W. W. Norton & Company.

Ogden, P., & Fisher, J. (2015). *Sensorimotor psychotherapy: Interventions for trauma and attachment (Csm)*. W. W. Norton & Company.

Porges, S. W. (2007). The polyvagal perspective. *Biological Psychology, 74*(2), 116–143. https://doi.org/10.1016/j.biopsycho.2006.06.009

Porges, S. W., & Dana, D. (2018). *Clinical applications of the polyvagal theory: The emergence of Polyvagal-Informed therapies (Norton series on interpersonal neurobiology)*. W. W. Norton & Company.

Riessman, C. K. (2005). Narrative analysis. In *Narrative, memory & everyday life* (pp. 1–7). University of Huddersfield.

Rogers, C. (1942). *Counseling and psychotherapy: Newer concepts in practice*. Houghton Mifflin. ISBN 978-1-4067-6087-3

Schore, A. N. (2012). *The science of the art of psychotherapy: The latest work from a pioneer in the study of the development* (1st ed.). W. W. Norton & Company.

Shulman, L. (2016). *The skills of helping individuals, families, groups, and communities*. Thomson Nelson.

Siegel, D. J. (2011). *Mindsight: The new science of personal transformation*. Bantam.

Siegel, D. J. (2021). *Becoming aware: A 21-day mindfulness program for reducing anxiety and cultivating calm*. Penguin Random House.

Sweeton, J. (2021). *Eight Key Brain Areas of Mental Health and Illness*. W. W. Norton & Company.

Van Der Kolk, B. A. (2014). *The body keeps the score: Brain, mind, and body in the healing of trauma*. Penguin Publishing Group.

Watson, L. B., DeBlaere, C., Langrehr, K. J., Zelaya, D. G., & Flores, M. J. (2016). The influence of multiple oppressions on women of color's experiences with insidious trauma. *Journal of Counseling Psychology, 63*(6), 656–667. https://doi.org/10.1037/cou0000165

Wood, K. M., & Geismar, L. L. (1989). *Families at risk: Treating the multiproblem family*. Human Sciences Press.

Chapter 8
Preliminary Assessment

1 Initial TICIBI Assessment

In social work practice, providing trauma-informed, culturally based intervention (TICBI) for survivors of wars with specific trauma responses begins with a collaborative gathering of information about the issues faced by the client and identifying their service needs. This involves recognizing their trauma responses according to the hierarchical three-part autonomic system or mindsets for maintaining human survival. Information gathering in social work hinges on connecting with the client, listening, observing, and asking respectful, culturally informed questions. To streamline the assessment and intervention processes, link your client's experiences with migration stages (see Chap. 2) and their trauma response (autonomic responses) to the three mindsets (see also Chap. 7 and the description below).

Survival mindset: This describes clients stuck in fight, flight, or freeze (an immobilizing behavior) that affects their motivation for change during the intervention. They deal with active sympathetic mobilization, causing them physical and mental challenges, including anxiety, fear, depression, insomnia, anger, and pain.

Therapeutic mindset: This describes clients showing some signs of the survival mindset but who are on their way to recovery. They exhibit emotional flexibility (seeing possibilities) and are open to discussing making a change. Multiple factors contribute to this openness to moving away from the survival mindset, requiring an in-depth assessment to support their recovery and advancement toward an active, engaged mindset.

Active, engaged mindset: This describes clients who are working on their recovery and moving toward growth and adaptation. They exhibit signs of an active ventral vagal system, sensing feelings of safety and increasing motivation to establish connections while actively participating in social engagement.

All three types of clients share a pressing crisis that led them to seek services, but they require varied intervention plans. Recognizing their trauma responses and

N. Bokore, *Trauma-Informed, Culturally Based Intervention*,
https://doi.org/10.1007/978-3-032-06835-4_8

service needs during the initial assessment is central to the success of TICBI. Trauma dysregulates the autonomic nervous system, so paying attention to your client's responses during this preliminary assessment allows you and your client to identify the current issues and the impact of trauma on their mental and physical health (e.g., depression, anxiety, fear, PTSD, chronic pain, emotional numbness, memory issues) and other crises affecting their lives. Table 8.1 illustrates the fundamental social work skills necessary to gather the required information for a comprehensive TICBI assessment.

Despite the varied mindsets of survivors, using your skill set in TICBI begins with your ability to be present and use attuned listening to hear your client's story.

Table 8.1 Activities during preliminary assessment

Activity		Skills used	
Exploring	**Attuned listening** • Ask for clarification as you summarize or paraphrase specific the conversation • Show empathy throughout the process • Listen for culturally based descriptions and show your understanding of resistance and change, which is often challenging	**Motivational interviewing** • Ask questions with empathy for further understanding, build confidence, and work on change • Identify trauma responses based on survivor's mindsets and the function of their autonomic system	**Use other interviewing methods** • Ask open-ended questions and listen to the stories within the survival narratives • Ask your client to elaborate on verbal and non-verbal actions based on your immediate observation, to identify autonomic reaction
Empowering/ Enabling	**Acknowledging strength** • Identify your client's values, motivation, capacity, and strength/ resilience • See the challenges as opportunities for PTG • Identify recourses • Reframe thought processes	**Information sharing** • Provide information about trauma and the hierarchical response of the autonomic system while answering your client's questions • Reassure your client about the benefits of the work ahead and the healing process	**Supporting** • Provide resources • Discuss referrals • Give positive and encouraging feedback about your client's resiliency and motivation for healing past wounds
Challenging	**Challenging** • Explore distorted thinking and provide informative feedback • Reframe thought processes • Explore values, capacity, and motivation and ask client to make a change	**Goal setting** • Identify the main issues and set measurable goals • Help clients explore alternative solutions and encourage them to work on change	**Taking responsibility** • Add to the conversation client's goals and their commitment to change • Follow the agenda of the intervention goals for planned change • Enforce the steps with empathy and flexibility to motivate clients to work on the planned change

Exploration is used to understand the issues, risks, and protective factors within your client's environment. As you use your skills, it is crucial to keep in mind the effect of trauma on the brain and how it affects your client's responses and increases their sensitivity to triggers. The exploration of your client's environment in this initial assessment is informed by person-centered theory. Developed by Carol Rogers in the 1940s, the person-centered theory has been used in the past 70 years by many practitioners to facilitate change. In trauma intervention, Rogers' (1951) work, such as having an "unconditional positive regard" for clients, helps solidify your initial connection, creating feelings of safety for clients who have lost this feeling during their journey of forced migration.

Unconditional positive regard creates a collaborative exploration. However, in TICBI, it is crucial to remember that some clients struggle to articulate their thoughts and feelings or with remembering the sequence of events because their trauma has affected their memory or created verbal amnesia. When providing TICBI in this and subsequent phases, patience is needed to decrease your client's stress and create a safe environment. Conducting the assessment in a quiet and clean space advances the reflective conversation about service needs. However, in TICBI, creating a safe space begins with you, the social worker, and your reflective preparation.

Safety in TICBI includes attention to gender-based social interaction, which may impact the relationship-building process. Considering your client's experience with trauma (e.g., gender-based violence, child abuse, and ethnic violence) and how it plays into safety or their fear is necessary. Another factor in this exploration phase is paying attention to dissociated feelings some survivors show because it affects their social interaction, including with a social worker. Dissociation, which can be a symptom of mental illness, is present in trauma survivors, manifested as difficulty making a connection between thoughts, memories, feelings, and behaviors (Ledoux, 1996; van Der Kolk, 2014). Therefore, in this exploratory phase, pay attention to the combination of challenges your client faced in the past, especially those related to emotions, feelings, and memories, for example, struggling with filling out multiple intake forms, knowing there are dire consequences such as denial of refugee status and services, if they fill them out incorrectly. Communicate in a friendly and empathic fashion to help them understand the change process as you begin the introductory information sharing.

1.1 Providing Introductory Information

Providing introductory information for TICBI is essential, starting with explaining the intervention process. Define your role and the type of service you provide (counseling or psychotherapy, often used interchangeably). The initial phases of TICBI can be threatening for survivors because they are usually asked to confront their past. Working with survivors with any of the three mindsets requires clear and straightforward language. Provide introductory information about your services and acknowledge that learning about the client, their cultural background, and their history pertinent to the TICBI process takes time.

Most resettled refugees in Canada come from non-Western nations. They are often unfamiliar with Western service provision, values, and the service language used in mental health intervention and psychotherapy. However, it is good to know that talk therapy is familiar to many survivors despite the different nomenclature. It is a culturally acceptable healing practice used in various cultures (seeking guidance or advice from elders or religious leaders) and is familiar to many. As part of the information exchange, providing information about the similarities and differences between what they are familiar with and what you offer is a good icebreaker for building collaboration. As you provide information, show empathy in the face of your client's concerns; understanding without judgment is a good start for this exploration phase and will strengthen your relationship. Using culturally informed communication throughout the assessment and subsequent intervention will engage your client. Establishing the intervention with this perspective ensures your client's safety and benefits the healing process with an added layer of co-regulation. Relationships are a biologically integrated human need that survivors seek despite their past painful interactions. Establishing a connection is essential for survival (Porges, 2017). In TICBI, it is necessary to understand how humans are biologically wired to connect. In the initial interaction with a social worker, a survivor's autonomic nervous system seeks a connection with the social worker, sending signals for a positive connection. Co-regulation in person-centered social work practice begins when positive signals meet a corresponding signal (Porges, 2017).

One caution in establishing a connection during this exploratory phase is understanding your client's cultural background—for example, sometimes family members will be involved in the process. This makes the co-regulation process multidirectional. In many communal cultures, family members are involved in health-related conversations, and your client's family members may be responding to an experience by changing the conversation from bi-directional interaction to multidirectional information gathering. When this happens, avoid Western personal privacy talk and continue gathering and providing information. This opens the door to discussing the process of TICBI and how the individual client and family members attending this initial session can provide valuable information and support. Continue with the TICBI by

- Introducing your qualifications and intervention approach. This is important because it helps your client and, if present, family members to build confidence in your skills and trust the intervention process.
- Explain how long each meeting will last, including this initial meeting, the maximum number of sessions, and the pros and cons of a shorter session versus an extended one.
- Explaining the possibility of having a follow-up session based on future needs.
- Give an example of the questions you will ask your client in this and other sessions. Also, give examples of what topics will be included in your questions and explanations, such as their experience with forced migration. Because questions

like this can be fear-inducing and possibly trigger your client, explaining the process and the pace of the intervention decreases the fear.
- Review the referral information to identify areas about which you need more information to provide TICBI. Start by collecting descriptive information.

1.2 Reviewing Other Descriptive Information

While reviewing the referral source(s), consider the agencies your client was involved with before being referred to you and the services they received. Establishing a connection, creating safety, identifying unmet service needs, and minimizing vulnerability, mistrust, and the sensation of danger are crucial components of the exploration. Thus, when reviewing the referral, pay attention to your client's sensitivity about questions about past services and current needs.

When collecting descriptive information

- Identify how your client prefers to be addressed (name, pronouns, e.g., she/her/they). Some cultures have strict rules regarding this.
- Ask about marital/partner status. This requires listening closely to your client's response to identify information about sexuality, which is a taboo subject in many cultures.
- Include your client's country of origin and forced migration, including the initial date, length of stay in transit countries before resettlement, and resettlement date.
- Identify how race, ethnicity, culture, and religion shape your client's worldview and the barriers impacting their capacity for change.
- Discuss the beliefs that drive your client's views as a way to explore their motivation for change. Spirituality is part of your client's meaning-making process. Involvement with religious organizations and practices allows you to explore community connections and identify resources, support systems, and opportunities for change.
- Explore the sources of your client's stress, including legal issues (e.g., fear of being caught or deported, problems with their residency status, court involvement, restraining order, or child support obligations).

1.3 Review Service Barriers

During this phase, past service barriers may impact the exploration stage for survivors, regardless of their mindset. For example, the needs of resettled refugees and immigrants are often systematically structured as one, but are fundamentally different, which can be confusing for clients and stressful for service providers. The differences between refugees and immigrants and their varying service needs are discussed in Chaps. 1 and 2. I remember the emotional stress I experienced when I was asked for documentation; the only way I could explain why I did not have it was by revisiting painful memories. I will tell my story of why I did not have certain documents for myself and my children, which affected my health and my chances of getting a job in the early days of my resettlement. When I was doing my PhD, I saw firsthand what happened to some of my community members who had similar circumstances and who became emotionally overwhelmed and gave up on the process, continuing to live in a cycle of trauma and stuck in the past because of their active dorsal vagal responses. Therefore, it is essential to pay attention to your client's emotional reactions to your exploratory questions and clearly explain the services available, including options for rebuilding their lives.

It is worth noting that other barriers may impact the intervention process, including language and the need for a translator. Even though translation is necessary to bridge the language and cultural gap, it can also create barriers:

- The translator may be from the same country and speak the same language as your client, but may belong to an opposing faction in the civil war that disrupted their lives. This may be a source of trauma for your client. Fear of discussing specific issues with the translator present creates a barrier to three-way communication. The translator may trigger painful memories (activating the neuroception of danger) and other issues impacting your client's connection with services and you, the social worker.
- Language barriers can impact your ability to communicate directly with your client, causing you to lose valuable information needed to provide TICBI.
- A challenge when using a translator who is a family member or friend is that they are aware of your client's discomfort in sharing sensitive community or family information. They may try to shield your client and refrain from translating what you say. Sometimes, your client may protect the translator from their horrific experiences.

Despite translators being necessary for providing services in some cases, bridging the linguistic and cultural divides, and being aware of the obstacles while collecting descriptive information sometimes requires creative ways of working with your client and a translator. I often used the conversation to draw emotional maps, such as genograms or culturagrams, designating spaces for the type of relationship, connection, and emotions they have in each map while using the social work skills listed in Table 8.1.

1.4 Provide Information About Confidentiality and Informed Consent

In providing TICBI, confidentiality issues are critical, as are your client's fears of discussing taboo subjects. Based on professional values, social workers believe in confidentiality and upholding the client's need for self-determination. However, confidentiality is affected by your client's culture and whether your services are mandated or self-referred. While a self-referred client has complete independence in the intervention process, you may need to define your working relationship differently for a court or agency-mandated client. Providing this explanation is essential for survivors of wars, allowing them to decide on topics and the process. The other challenge of confidentiality is culture and the involvement of family members in the process, making it challenging to discuss privacy. To avoid this situation, explain what confidentiality means and your concerns about confidentiality, including your agency's policies and privacy parameters. When working with survivors in TICBI, please do not rush the discussion; remember how traumatic memories shut down the cognitive process or, for some, cause verbal amnesia, preventing them from putting their thoughts into words. Give your client time to consider your information and answer your questions. Your openness to this form of understanding and flexibility strengthens your relationship with the survivor, which is the basis for providing TICBI. Use simple words and language to avoid overwhelming your client with scientific or technical terms as you explore your client's background using a bio-psychosocial, structural, and cultural assessment.

2 The Bio-Psychosocial, Structural, and Cultural Assessment

2.1 Explore Biological Factors

Without precise chronological data showing their history and vulnerability to the activation of traumatic memories, survivors continue to either adopt a survival mindset or fluctuate between different mental states, maintaining a chronic sense of danger characterized by intrusive fear, feelings of dread, hypervigilance, body memories, and confusing thoughts and behaviors. Survivors often seek services when grappling with these usually unexplained feelings and behaviors, which can include constant anxiety, depressive moods, shame, low self-esteem and feelings of loneliness or alienation, problems with anger, or chronic impulsivity that affects their social relationships. For effective TICBI, include biological factors in your assessment to identify your client's emotional responses to real or perceived threats that induce these stress-inducing conditions. Pay attention as survivors recount their stressful stories about what triggers stress and activates their hypothalamic-pituitary-adrenal axis, leading to increased cortisol production. As you explore your client's daily life stressors and social interactions, be mindful of how they describe emotions and the coping mechanisms they employ to regulate their reactions or modify their behavior.

Since adaptation to survival is triggered automatically in the face of external threat, exploring their emotions and physical sensations or implicit reactions is necessary. Structure your conversations so they invite stories about fear-induced reactions (fight, flight, or play-dead) to help you understand their implicit responses.

This opens an opportunity to discuss how these experiences affected your client's well-being: for example, these fear-based reactions create a neurochemical chain reaction activating their amygdala (smoke/danger detector) and the hypothalamus, which takes a central role in initiating the adrenaline release in the sympathetic nervous system initiating the fight or flight response. This process increases the release of adrenaline and speeds up the heart rate and breathing to increase the oxygen intake to support muscle tissue and prepare the body for escape from danger. Unfortunately for survivors, the chronic production of adrenaline has adverse effects on their bodies, including hypertension, a heart attack, or a stroke. An everyday complaint your client might be dealing with is weight gain because of overeating to suppress painful emotions. Since stress is part of their daily life as they experience forced migration, persistent surges of adrenaline mean survivors continue to deal with its effects on their bodies. As part of your practice

- Explore the causes of emotional dysregulation, which triggers a neurochemical chain reaction in your client's life and challenges their body and brain responses, exacerbating their mental and physical distress (causing constant hyperarousal and hypervigilance)
- Listen to your clients' stories and how they explain their fears, worries, anxieties, and restlessness. Look for connections, such as hearing bad news from their home country or dealing with stressful or dangerous situations in their new home, including neighborhood violence and death threats.
- Listen to the stories about their pre- and post-migration physical health, focusing on post-migration complaints such as chronic pain, gynecological issues if your client is born female, digestive problems, asthma, accelerated heart rate, and joint and muscle pain

There is a shortage of literature exploring the mental health issues and health challenges faced by resettled refugees, as well as the barriers they encounter as a result. As a survivor who developed physical impairments during torture, I struggled with a lack of access to information and services when I initially arrived in Canada, which could have made a difference in my life. At that time, there was no easy online access or social media to find information upon my arrival. I relied on service providers' willingness to share information or on personal searches, which were not easily accessible for someone new to the country. For those dealing with disabilities, the lack of information and resources increases family stress. These stressors are compounded when there is a culturally based stigma attached to specific disabilities, such as mental illness. Therefore,

- Explore disability, access to resources, and support systems or services at home for those who need them in their neighborhood or the workplace.
- Assess whether a family member is experiencing disability challenges or has care needs, and recommend available services to assist them.
- In some cultures, disability is stigmatized and hidden. Explore your client's views on these issues and their cultural context.

- Explore your client's use of medication in the past or present, new or old eating habits, sleep patterns, and taboo subjects affecting their health and relationships.

2.2 *Explore Psychological Factors*

As previously mentioned, trauma is an invisible wound with deep emotional scars that can lead to behavioral stuck points, resulting in autonomic dysregulation and stress responses. The *Diagnostic and Statistical Manual* (DSM-5) provides a list of PTSD symptoms that you can use to initiate exploratory and educational conversations. Depending on your client's culture, some psychological responses may exacerbate their fears, heightening anxiety or causing panic attacks. These reactions may lead them to withdraw from therapy or develop a detached attitude toward the process. Experiencing trauma creates a complicated psychological response, some aspects of which can interrupt the intervention process. To explore the psychological impacts of trauma, consider using some of the interactive questionnaires listed in Table 8.2. These questions serve as preliminary tools that can help identify the presenting issues. It is important to remember that experiencing trauma and addressing traumatic wounds can affect personality and social interaction due to the presence of emotional dysfunction or stuck points influencing the client's thought process. Allow your client time to become familiar with or adapt to these

Table 8.2 TICBI self-evaluation questionnaire

Emotional arousal and body/brain response	Response
Are you often preoccupied with what happened in the past? Do these thoughts worry you?	Yes/No
Do you often think about the past, seeing flashback images of your experiences?	Yes/No
Do these feelings, images, and memories make you sad or anxious?	Yes/No
When you think about a time when you felt sad or anxious, did you notice any related physical responses or sensations?	Yes/No
When you think about or discuss the past with others, what are the physical sensations you felt? In those cases does your body feel tight or do you experience joint or muscle pain?	Yes/No
Do you have thoughts and emotions you cannot identify where they are coming from but make you feel cold, helpless, and sometimes angry?	Yes/No
Do you sometimes have difficulty concentrating or focusing on the task at hand?	Yes/No
Do you often repeat the same story, forgetting if you've told the story to the same person before?	Yes/No
Do family and friends often tell you that you are stuck in the past?	Yes/No
Do you often focus on what is wrong in your life and forget the good moments?	Yes/No
Do your family members and others often remind you to move on and think about positive things, such as resettlement and adaptation-related activities?	Yes/No

intervention tools, inviting them to take them home and bring them back for discussion in the next session.

Tally up the "yeses" and "no's" to discuss your client's psychological responses and stuck points that must be part of the intervention process. Ask questions to understand the presenting issue(s) further and continue with the change talk. Adding simple, user-friendly, and straightforward questions based on your client's situation (and experiences) is essential to enhance the discussion. Additionally, when discussing these self-evaluation responses, you can continue to incorporate self-reflective questions to raise awareness about the past and its psychological impact on your client. Having follow-up or after-activity feedback is a helpful encouragement for future participation and change. Here are some examples of follow-up questions you might want to ask:

- What do you think about this question?
- What does your higher score of yeses tell you, and were you surprised by it?
- Regarding the yes responses, which area should we focus on in our collaborative change talk?

As you discuss these questions and the scores (yes/no), observe your client's physical responses and assess their sympathetic responses and dorsal vagal activation or collapse (hyper-freeze or hypo-freeze): faster breathing, increased heart rate, or muscle tension. This allows for further exploration to prepare your client for additional work and the intervention, during which you will discuss issues they are struggling with and healing practices. Starting a discussion with these self-evaluation questions helps you identify hidden concerns as you pay attention to what triggers hyperarousal or hypoarousal in your client, thus narrowing their window of tolerance.

Self-assessment questions also help you explore the reasons behind your client's emotional shutdown and fight-or-flight, or dorsal vagal, responses. For survivors, recalling the past may evoke overwhelming and frightening emotions, such as a sense of danger or hopelessness. Be cautious not to push your client into uncomfortable feelings that could trigger their adaptive and self-protective behaviors, which might halt the intervention or cause them to drop out of treatment.

Be patient as you engage in a psychological assessment during the exploration phase. As previously mentioned, traumatized individuals often fixate on specific events from the past, which keep them entrapped and cause constant emotional dysregulation. When that happens, pay close attention to the emotions related to the questions in Table 8.1 and the activation of specific brain areas such as the amygdala, where fearful or threatening stimuli are processed, leading to sympathetic mobilization and a shutdown of the prefrontal cortex, which affects the management of thoughts, actions, and emotions.

In this process, the vagal brake, also known as the heart's pacemaker, is a regulatory action linked to the ventral vagal system, managing sympathetic and parasympathetic mobilization that controls heart rate and overall autonomic responses. In TICBI, exploring survivors' emotions (engaging or disengaging the ventral vagal system) is essential for evaluating how they manage day-to-day life stressors based on their past adaptive responses, including early life experiences that helped them develop self-regulation and functional flexibility of their vagal brake (Cozolino, 2021; Porges, 2017). A well-functioning vagal brake is a source of internal resilience and post-traumatic growth. It grants survivors a unique ability to quickly engage in survival actions and disengage when necessary, creating calm and allowing them to reorganize or envision life in post-traumatic circumstances (Dana, 2020).

The vagal brake can be activated through various techniques designed to stimulate the vagus nerve (see Chap. 10). A well-functioning vagal brake helps move survivors from a survival mindset toward an active, engaged mindset and the possibility of post-traumatic growth. TICBI activities in Chap. 10 aim to strengthen the vagal brake by promoting learning about self-regulation and developing methods for self-understanding and controlling behaviors, managing reactions, and actions in stressful and fearful situations. During the exploration and later intervention phases, you and your client must rely on trust and the strength of your relationship, facilitating discussions about self-regulation and managing the triggers of past wounds. During the psychological exploration phase, co-regulation guides the conversation to uncover additional information needed for the intervention, such as the social factors creating barriers to establishing social connections.

2.3 *Explore the Social Factors and Barriers*

For resettled refugees, the initial years—decades for some—come with resettlement struggles: lack of a well-paying job, language barriers, and lack of resources and housing. This is even more challenging for survivors of wars who lack social connections, which is strongly linked to parasympathetic nervous system activation, contributing to survivors' overall well-being. It keeps survivors away from their fight, flight, and freeze responses (sympathetic or dorsal vagal activation), moving them toward ventral vagal activation. In light of this, explore your client's strengths, including resources and the presence or absence of social connection:

- Explore barriers to job opportunities, such as your client's education level (both pre- and post-migration) and whether they had documentation upon arrival. Some resettled refugees, like me, despite having advanced education, find themselves continually unemployed or underemployed due to systemic barriers, a lack of social connections during the initial phases of resettlement, and the intersectionality of oppression (see Chaps. 1–4).
- Explore barriers to establishing connections or getting support from friends, community, religion, and work.
- Explore barriers to adaptation (e.g., psychological adjustments, difficulty learning a new language, adjusting to a new culture, identity crisis as part of a minority group, or the challenges in dealing with the intersectionality of oppression.

- Explore the impact of delayed grief (suppressed or postponed during stages of forced migration). Delayed grief is common among refugees who deal with multiple losses during their forced migration and can be a barrier to socialization.
- Since some resettled refugees find themselves isolated (by language barriers, lack of information, and fear), explore accessibility to culturally or religiously appropriate foods (halal, kosher, vegan), places of worship, and community organizations.
- Explore health issues that contribute to your client's isolation and impact their well-being.
- Explore other isolating factors, such as taboo-related issues, fear of exposure, and the need to stay hidden. Since survivors experience various forms of violence during their displacement, the impacts of those experiences keep some stuck in spaces of survival (see Chap. 3).

2.4 Explore Structural Factors

As discussed in Chaps. 1 and 2, systemic and structural issues in their home countries, including political factors, forced survivors to leave their homeland. Cautiously explore the political context of your client's past, contributing to their struggle to establish trust or share information with you and build a collaborative relationship with others in their present environment. Proceed cautiously during this exploration session to avoid triggers, using all the skills outlined in Table 8.1 to understand their resistance to sharing painful stories hidden within their survival narrative. It is also important to remember that the resistance presented by survivors, particularly those of marginalized groups, is linked to existing systemic and structural factors. Thus, proceed cautiously, paying attention to verbal and physical responses during this phase, and provide information and resources that are encouraging and available to them.

2.5 Explore Risks and Vulnerabilities

For some survivors of prolonged trauma, bio-psychological and social factors increase their self-isolation. In these challenging times, when inflation is high while the social welfare benefits for resettled refugees have plateaued, homelessness is common among resettled refugees. This is because of the high cost of housing and the financial difficulties brought on by unemployment or underemployment. Vulnerabilities for survivors increase as they are pushed into low-end, crime-ridden neighborhoods, which trigger memories of past experiences. Explore these and other safety concerns in your client's living situation that compound their stress and increase their vulnerability.

Survivors with any of the three mindsets, but particularly those with a survival mindset, may be dealing with personal vulnerabilities, including self-harm, hidden addiction, self-neglect, and family violence. Pay attention to cultural changes, including expressions of sexual identity and family violence. Cultural restrictions induce fear and isolation from family, friends, and community; survivors may keep specific issues hidden, but this adds to stress and damages health and well-being. As you explore these issues, consider diversity within diversity and the barriers created by the intersectionality of multiple traumas. As mentioned in Chap. 6, survivors, mainly those with a survival mindset, turn to self-destructive behaviors, self-medicating practices (traditional or otherwise), and substance use, such as khat, alcohol, and cannabis.

Utilizing empathetic motivational interviewing (Rogers, 1942) is essential for building rapport, facilitating change, and fostering a collaborative relationship. As you pose the question, demonstrate unconditional positive regard for your client and their experiences, showcasing acceptance and respect for their perspective as reflected in their survival narratives. TICBI allows us to view survivors' perspectives with dignity, believing that every culture has inherently good and protective aspects that help members to live in harmony, collaborate, and support each other, allowing its members to achieve their potential. Therefore, it is important not to judge other cultures and perspectives, even when disagreeing with a cultural explanation of specific issues or taboo subjects. The success of TICBI is achieved through collaboration, trust, and understanding when sharing information and resources. Providing services with these values contributes to healthy collaboration and working toward change by giving survivors a chance for reflection, resiliency growth, and healing (see Rogers, 1951).

2.6 Explore Spirituality, Religiosity, and Resiliency

When working with diverse groups, it becomes evident how many of them see spirituality and religion as essential to overcoming stressful life events. They play a part in intervention programs and are often vital in healing. In the exploration stage, they can open the door for a conversation about your client's belief system and how it creates strength and helps them deal with issues.

While exploring the role of spirituality and religion in healing, it is necessary to differentiate between them. For example, religion is often described as an institutional framework organized around a belief system, traditions, rituals, and sacred texts. In contrast, spirituality is seen as an interpersonal matter, and people seek meaning, purpose, and connection with a higher power to direct their lives. Survivors often define this as using spiritual practices to help them, such as meditation or mindfulness.

Although the terms *spirituality* and *religion* are often used interchangeably, in TICBI, they are distinct. Both religious practices and inner spirituality shape a survivor's sense of meaning and purpose, influencing their recovery and healing process. My strong sense of survival and religious beliefs shielded me from many painful experiences and contributed to my strength, resilience, and hope. My flexibility in adapting to healing practices such as meditation and yoga helped me manage emotional and physical pain. Each practice provides survivors with value-based strategies for coping with adversities, exploring positive contributions while discussing and building strength and coping mechanisms. Therefore, during the assessment and exploration phase of TICBI, evaluate how often your client seeks spiritual and religious support when facing adversities (e.g., prayer for strength, safety, and forgiveness).

Caution is needed when exploring spirituality and religion, as they can also be sources of pain, guilt, internal conflict, and alienation for some survivors. Therefore, explore the specificity of the restrictive religious and spiritual practices that create fear and affect your client's resettlement and socialization, but remain open to their restorative powers.

In social work, providing TICBI while considering culture, religion, and spirituality also means keeping in mind the personal strength and resiliency that help survivors withstand trauma. Since past strengths predict future actions, ask questions about what enabled them to overcome challenges similar to the ones they are facing in the present. What is one resilience-based quality that helped them to survive in the past? Finally, asking additional questions to explore personal strength, resiliency, and problem-solving abilities, including searching for available resources and support systems, helps identify the presenting issues and set goals for change.

3 The Presenting Issues

Based on the information you gathered during the exploration stage, you can now identify the main issue or concern your client needs help with. Carefully assess the collected data to formulate the shape of your client's program. A collaborative review of the data gathered helps develop an initial hypothesis about service needs. It identifies what makes specific issues persist, keeping your client in a cycle of trauma (survival mindset).

As mentioned in Chaps. 3 and 7, survivors function within their autonomic response, moving up and down the ladder of the vagal state (ventral vagal, sympathetic, and dorsal vagal), which determines their mindset. Identifying the problem effectively depends on the quality of your relationship with your client. In this process, employing a strengths perspective (Saleebey, 2013) helps establish trust, which can often be challenging for survivors who have experienced a breach of trust and unsafe conditions.

3.1 *Identifying Issues from a Strength Perspective*

The strength perspective empowers survivors to work toward growth by developing skills they can use to move forward and heal. It begins when you collaboratively identify the presenting issues, making your client feel, for the first time, that they have been heard and understood and possess the strength and power to change their lives. Utilizing a strength perspective allows for further discussion to take place. It is essential to remember that while trauma is a wound that leaves a scar, the cause of your client's latest crisis or service need can also become an opportunity for post-traumatic growth and making a change for a better life (see Chap. 4). Understanding your client's adaptive sympathetic or dorsal vagal mobilization and its impact on their mindset is important. This understanding will help address barriers that prevent an activated or halting ventral vagal response (see Chap. 7—"The Three Survivor Mindsets"). To offer further psychoeducation, your training and additional readings are necessary to create hope, broadening your client's perspective on their problems so they can envision healing.

3.2 *Providing Psychoeducation About the Presenting Issue(s)*

Survivors of prolonged trauma often cannot articulate the significant impact it has on their lives because of loss of memory and barriers created by their adaptive reactions. Psychoeducation is the core of TICBI, designed to provide survivors with culturally informed, relevant, and respectful information about their past traumatic experiences, their impact on body/brain, and the recovery process. As discussed in Chap. 3, trauma can impair cognitive flexibility, affecting survivors' thinking and decision-making processes. Survivors seeking services may be dealing with the disruption of connectivity between brain regions involved in stress response and executive functioning. This is particularly evident in those with a survival mindset who are struggling with PTSD symptoms. While exploring the survival-based behaviors that have become part of your client's personality, define their psychosocial functioning based on the data you collected during assessment, including their adaptive responses to life events and personal relationships. As you review relevant referrals from medical doctors, schools, and family members, ask questions and explain those responses, focusing on the impact of past experiences on their well-being.

For some survivors, fear and adaptive survival responses create mental rigidity that resists change. Paying close attention to this, particularly when working with clients in a survival mindset who have experienced chronic trauma and fear, is crucial, as they often lack the emotional flexibility and reflective thinking necessary to make a change. This rigidity leads to chronic isolation and alienation, significantly limiting the social support needed for growth and well-being. Sharing information, or psychoeducation, helps explain the impact of cognitive rigidity and inflexibility (seeing the world in black and white) on your client's life and the benefits of practicing cognitive flexibility.

Offering information about how social isolation affects mental well-being and activates fear responses, how self-protective and adaptive behavior contributes to further isolation, and how self-estrangement results in living with emotional despair solidifies the intervention process and underscores the need for change. For individuals from communal societies like mine, a trauma response manifests as the development of self-isolating behaviors to avoid being perceived as someone who struggles with health and mental health issues. For others, self-isolation is not a choice but a behavioral adaptation to cope with the loss of collective support and the lack of emotional connection with new communities due to language barriers and cultural differences. These behaviors further perpetuate isolation by continuing the activation of the fear circuit (amygdala) and/or dorsal vagal activation. In the past few decades, I have witnessed the fear and suffering in silence among some community members increasing due to their loss of connection and communal social support systems.

As you move on to setting intervention goals, use simple examples to communicate and share knowledge about these issues, providing a neurological response to trauma for increased self-understanding and stabilized feelings. Frame the information in culturally informed language to increase your client's self-confidence and motivate them to continue working toward change and healing. Psychoeducation decreases the stress of fear related to challenging conversations, including cultural taboos, which often overwhelm survivors' emotions and affect their well-being.

4 The Challenges of Psychoeducation in Social Work

It is important to remember that many survivors may not understand the technical, scientific terminology used to describe trauma and body/brain responses. Therefore, during the assessment or exploration phase, it is crucial to consider language use, particularly in how to share knowledge in a culturally responsive way (awareness of the values and meaning-making process of others or clients from diverse backgrounds). For example, the challenge of sharing knowledge about brain/body reactions to trauma using neuroscience requires translating simple words for culturally responsive learning.

Despite this challenge, assessing the impact of trauma using scientific knowledge about changes and reactions after trauma proves helpful. Other factors to consider in this process include awareness of the client's capacity to understand the information based on their level of education and the impact of trauma. Evidence-based knowledge helps normalize clients' struggles and instills hope for recovery and healing. It is also important at this stage to remain mindful of some survivors, particularly those stuck in a survival mindset, as they may contend with stress-induced memory loss affecting the hippocampus (the memory center), which impacts information exchange (see Chap. 3). This stress-induced memory loss, resulting in dissociative amnesia, has been linked to PTSD due to the disruption of the hippocampus's normal functioning, which is necessary for memory and the ability to recall past events. This issue may become evident during assessment, exploration, or talk therapy due to the stress the client might experience or the client's struggle with the disconnect between their thoughts about past events and their ability to articulate them.

In TICBI, the assessment and exploration phase also includes sharing information about top-down and bottom-up (cortical to subcortical and back again) and left-right (across the two halves of the cortex) processes, which help to enhance the healing process. Psychoeducation, a simplified method of sharing information about trauma responses, involves discussing how these responses originate in the old brain, specifically in the brain stem and limbic system (bottom-up), and are designed to trigger rapid actions for survival and sustaining life. The functions of the reptilian brain, known as the brain stem, manage the survivor's instinctual feelings and sensations that govern basic survival actions and reactions, including physiological activities such as breathing, maintaining body temperature, and sleep cycles. Therefore, it is beneficial to discuss the limbic area and how it regulates or integrates critical mental and emotional adaptations learned during early years and used throughout life.

Similarly, the slow-functioning cognitive process (top-down), part of the newest brain, often shuts down first and only later engages in survival mode (when the danger has passed) to find solutions, including participating in the change talk. Discussing the function of the neocortex, the latest part of the human brain, and how it manages survivors' thinking, cognitive processes, and perceptions, including how they interpret their experiences, is a thought-provoking conversation to have with clients. To simplify knowledge sharing in this phase, consider making psychoeducation a brain talk, focusing on the healing aspect of the collaborative work that survivors can anticipate as they view their trauma wounds (previously challenging experiences) with self-empathy and self-understanding and look forward to their healing journey. In contrast, they work on their collaboratively set goals.

As you prepare to set goals for TICBI, provide hopeful information that discusses the benefits of change talk, or the talking cure. Your client's strengths and motivation to participate are integral to encouraging them to adhere to the treatment plan. Offering an opportunity to share survival narratives filled with hope at their own pace helps stimulate the Broca area and alleviates anxiety associated with the initial phases of an intervention. During goal setting, avoiding self-defeating language and using words of hope can motivate survivors and restore hope and healing, initiating change through neuroplasticity (Cozolino, 2010, p. 307).

5 Long- and Short-Term Goals

Setting goals depends on understanding the issues needing attention, your client's mindset, motivation, values, and the commitment to work on the planned change. In TICBI, goal setting involves using the activities outlined in Table 8.3, which facilitate reflection on the benefits of successful change on well-being and what happens when there is no change plan or action. These activities often motivate clients to adopt a change path, even when they face multiple barriers.

For many trauma survivors, particularly those with a survival mindset, the past is continuously remembered implicitly, often felt as real and sometimes making it difficult to verbalize those challenges, which keeps some survivors stuck in the past. Breaking the cycle requires actionable goals to change automatic beliefs and habitual patterns at each step. Since change is always tricky, setting goals necessitates considering the person-in-environment approach and organizing the intervention collaboratively. In TICBI, designing a change at critical points involves clarifying and organizing each step. Divide the work into manageable tasks to make it achievable.

For example, discuss the goals needed to address immediate needs. For someone with a survival mindset, this might involve finding shelter, food, school, and medical care (short-term goals). These goals are rooted in immediate needs and available resources and can be achieved quickly. Alternatively, it may involve reducing the habitual sympathetic actions adopted for survival (fight, flight, or freeze) (long-term goals). Living in a chronic state of survival (constant activation of the neuroception of danger) keeps individuals in crisis mode, negatively impacting their well-being. Change talk requires time for some survivors; therefore, they must pay attention to triggers and work on actionable steps that foster adaptive change (moving from a survival mindset to a ventral vagal state and an active, engaged mindset). The process of co-regulation, which includes mindfulness practices or discussions about acceptance based on values (spirituality or religion), is fundamental, along with constructive feedback aimed at motivating your client at each small step toward change and healing. It is essential to assess progress, evaluate outcomes, and celebrate success. In TICBI, using a simplified version of what is known as SMART goals, demonstrated in Table 8.4 (Doran, 1981; Egan, 2010),[1] can help survivors establish clear intervention goals without overwhelming or triggering them.

Table 8.3 Activities to explore client's planned change

What is the benefit of making a change to solve my immediate concerns?	
List the short-term benefits	List the long-term benefits
What is the impact of NOT making a change to solve my immediate concerns?	
List the short-term impact	List the long-term impact

[1] The goal-setting framework was popularized by George T. Doran in 1981 defining the creation of "Specific, Measurable, Attainable, Realistic and Timely" manners of managing goals. Since then SMART is used by various disciplines.

Table 8.4 TICBI SMART goals

	SMART Goals
S	**Specific:** In TICBI, clarity and specificity are significant in making a change. As much as possible, use simplified writing and cultural clarity. Explain each action using calming and motivating language for co-regulation; use motivational interviewing to determine what steps your client will take to reach your goals
M	**Measurable:** In TICBI, measurable steps include changing adaptive responses based on sympathetic activation or dorsal vagal mobilization activation to actions showing ventral vagal activation (e.g., self-isolation to increased socialization). To motivate your client to take actionable steps, use motivational interviewing to ask how your client will identify when their goal is met
A	**Achievable**: Avoid a triggering conversation at this stage and celebrate small or significant accomplishments for clients moving forward. Survivors struggle at times with fluctuating emotions (moving forward but they face a setback and backward; see the case study for Sarah and Nimali; see Chap. 9). In this discussion, emphasize body–brain fear sensations, measured on a scale of 1–10, ten being the highest and one being the lowest. As you use this scaling fear-response activity, pay attention to moments when your client benefits from co-regulation, moving toward ventral vagal activation (socialization and connection), and discuss it with them as a motivation and offering feedback
R	**Relevant and realistic**: For traumatized individuals with all three mindsets (see Chap. 9), triggers are familiar. As is the experience of being stuck in a cycle of emotions and failure to overcome it. Therefore, set realistic goals, explain both the successes and setbacks, explain the benefits of activating their ventral vagal system in these activities, increase their parasympathetic response through mindfulness activities, and pay attention to your client's emotional and physical sensations. It is necessary to use motivational interviewing to get at the values linked to your client's motivation for change
T	**Timely:** Create an explicit schedule for achieving the goals. Build in specific time frames, targets, and due dates to measure the relevancy of the goals and your client's achievement of them. If your client is struggling at some steps, re-evaluate the SMART goals, making the necessary modifications to motivate your client. Set a specific timeline and have your client make a commitment: "I will complete this task by (date)"

5.1 SMART Goals For TICBI

When working with survivors of prolonged trauma, summarizing the activities in this phase and specifying the work needed to establish specific, measurable, and timely goals based on your client's mindset is necessary for providing TICBI. Similarly, address safety issues early on while discussing past and present coping strategies and available support systems, including the advocacy needed for your client to access the needed resources. TICBI requires readiness (see Chap. 7) to work from crisis perspectives to minimize the impact of triggers from prolonged trauma. Goal setting includes working toward change, which is daunting, and hope-based conversations create opportunities to benefit from co-regulation.

For survivors with a survival mindset and, occasionally, for those with a therapeutic mindset, moving forward is not a straightforward action; those with an active, engaged mindset will, from time to time, revert to their earlier survival responses because of environmental stressors. Normalizing your client's current trauma response through psychoeducation is the key to helping them regain control and continue working on their goals for change. Use a process of self-checking on progress.

5.2 Checking Client's Feelings About the Process

Before moving on to the next phase of TICBI, the intervention, have a self-checking conversation with your client about the exploration process, identifying issues, and modifying goals if needed for a planned change. You can explain the process using sentences like, "Before moving on to the change talk, I will ask a few process-checking questions to determine your feelings about the exploration/assessment and goal-setting steps. When I ask you the question, please take your time and reflect on your feelings before answering."

Questions to ask:

- What has changed for you since we began gathering information about your experiences?
- Are there any issues or information you have not mentioned that might be useful for setting goals for change?
- Some of the issues we discussed may have been stressful for you. As you reflect on them, are there issues you avoided because of stress and fear? Why?
- While discussing the issues, are there any positive experiences or resources you recognized that would be helpful in your journey toward change and healing?

5.3 Checking Social Worker's Feelings and Practice Skills

As discussed in Chap. 7, TICBI is a collaborative process where the client and the social worker work together to find a solution. In that process, social workers are impacted by consistently hearing stories of extreme life challenges. They are at risk of bringing out their past trauma or unconscious bias about the issues or the person they are working with. These are environmental conditions for the client that they might not be aware of, but implicitly act upon. Reflecting on the following questions before starting the treatment phase of TICBI will help social workers recognize the presence of such issues and take steps to address them. Identifying the challenges (if any) at this early intervention stage and seeking support, training, or supervision is essential. The self-checking process also prevents social workers from miscommunicating, misinterpreting, and misdiagnosing, which decreases their stress level and health risks, mitigating burnout and vicarious trauma (Chap. 5). Here are a few self-checking questions you can use when uncomfortable feelings arise in TICBI.

- Did you start or end this session with thoughts of not wanting to continue working with this client?

- In the past, have you canceled appointments or failed to recognize when you felt uncomfortable about a client, the topic of the intervention, and the impact this had on your client?
- Did you judge this or clients (resettled refugees) by comparing them to other immigrants?
- Do you need help identifying or acknowledging how your client's issues might be attributed to their environment?
- Do you see yourself as an advocate committed to working for social and economic justice or changing conditions that push individuals, families, and communities into further trauma and persistent poverty?

Another self-check social workers can do at this stage is to take 15 minutes at the end of a session to reflect on it. Ask yourself,

- What are the issues that had an emotional impact on me?
- As the client shared their story, what parts of the exploration phase affected me the most, and how will that affect my future work with this client? You may want to consider your feelings and reactions.
- What challenges can I anticipate while working with this client?
- What are the strengths and resiliency factors about the client that captured my attention?
- Are there skill sets I need to improve before starting this client's treatment plan?

Definitions

Autonomic nervous system (ANC): The autonomic nervous system regulates physiological responses to the environment. It is part of the peripheral nervous system.

Bio-psychosocial assessment systematically explores the complex biological, psychological, and social factors and environmental interactions in a survivor's life that affect their internal responses and well-being. In TICBI, a holistic approach is used to assess the 4Ps (pre-disposing, precipitating, perpetuating, and protecting) in a client's life to understand what triggers their trauma or creates physical or mental health issues.

Case conceptualization: In TICBI, it refers to the explanation of a client's presenting problems.

Case formulation: The process of case conceptualization and creating a hypothesis about the presenting problems.

Co-regulation: When the nervous systems of two individuals interact to support each other and emotionally connect through mutual regulation of their physiological states.

Cultural assessment: Asking about ethnic identity, religious and spiritual practices, tribal and family rules, and health practices that function as precipitating or protective factors to understand clients' issues.

Epigenetics: The study of how behavior and environment influence gene expression during the prenatal period and early years. These experiences do not alter the DNA sequence but modify how the body interprets a DNA sequence. Unlike genetic mutations, these changes are reversible through intervention.

Goal setting: A structured plan to motivate and guide toward desired outcomes. In TICBI, it enhances survivors' engagement and collaboration with the steps of change talk for recovery and healing. It is a collaborative guide for social workers and their clients, providing accountability and motivation. Goal setting for TICBI also defines who is responsible (the client or the social worker) for achieving timely and measurable outcomes.

Self-estrangement: The experience of losing a sense of self when feeling emotionally and spiritually disconnected from others.

Parasympathetic mobilization: A neural mechanism significantly alters heart rate during engagement or disengagement in a conversation or encounter. It fosters self-soothing behaviors for emotional calm.

Psychoeducation: Offering educational information to understand the aspects of trauma, individual reactions, and related treatments.

Structural assessment: It uses critical insight to connect personal struggles with systemic and structural barriers. This view considers inequality a social justice issue and a health determinant. For TICBI, this also explores the political, economic, social, and environmental factors that create systems leading to displacement and disparities among survivors (e.g., racism, sexism, ableism), resulting in systemic violence, exploitation, and poverty.

Vagal brake: A neural mechanism described in the "Polyvagal Theory" framework that regulates heart rate during high levels of glucose and adrenaline (stress) and in calm states, enabling the individual to engage or disengage from their environment.

Discussion Questions

1. Think of a time when you, a family member, or a friend went through a traumatic event and sought assistance from services. What did the assessment and exploration process look like during your initial meeting with a service provider? Which aspects of the service(s) or the provider's skills made you feel safe, heard, and connected?
2. As you read this chapter, do any skills discussed resonate with your field experience?
3. List various challenges you may face when providing TICBI for clients with each mindset. What is your solution for those challenges?

Review Points

1. Explain why collecting information about your client's health, culture, religion, and spirituality is necessary during the exploration phase of TICBI.
2. The brain's smoke alarm system, or fear circuit activation, was discussed in Chaps. 3 and 4. Discuss how it could be a factor to consider during the exploration phase.
3. Similarly, the relationship among the amygdala, hypothalamus, and pituitary glands, collectively known as the stress pathway or the hypothalamic-pituitary-adrenal (HPA) axis, was discussed in Chaps. 3 and 4. How will you apply that information during the exploratory phase?

Competing Interests The author has no conflicts of interest to declare that are relevant to the content of this chapter.

References

Cozolino, L. J. (2010). *The Neuroscience of Psychotherapy: Healing the Social Brain (Second Edition)*. W. W. Norton & Company.

Cozolino, L. (2021). *The development of a therapist: Healing others - healing self.* W. W. Norton & Company.

Dana, D. (2020). *Polyvagal exercises for therapists and clients: 50 Client-Centred Practices*. W. W. Norton & Co.

Doran, G. (1981). There's a SMART way to write management's goals and objectives. *Management Review (Saranac Lake, New York), 70*, 35–36.

Egan, G. (2010). *The skilled helper* (9th ed.). Brooks /Cole.

Ledoux, J. (1996). *The emotional brain: The mysterious underpinnings of emotional life*. Simon and Schuster.

Porges, S. W. (2017). *The pocket guide to the polyvagal Theory: The transformative power of feeling safe*. W.W. Norton.

Rogers, C. (1942). *Counseling and psychotherapy: Newer concepts in practice*. Houghton Mifflin. ISBN 978-1-4067-6087-3

Rogers, C. (1951). *Client-centered therapy: Its current practice, implications and theory*. Constable. ISBN 978-1-84119-840-8

Saleebey, D. (2013). *The strengths perspective in social work practice (6th ed)*.. Pearson.

van Der Kolk, B. A. (2014). *The body keeps the score: Brain, mind, and body in the healing of trauma*. Penguin Publishing Group.

Part V
The Change Phases of TICBI

Chapter 9
Intervention, Evaluation, and Termination Phases

1 Introduction

After completing the TICBI-based assessment and goal-setting phase (Chap. 8), the intervention focuses on establishing goals prioritized as short-term and long-term. Providing an intervention involves sharing information communicated in culturally informed language to enhance awareness of trauma and its physical, mental, and behavioral impacts. This awareness allows the social worker to normalize the trauma responses that their client has been experiencing, such as sleep disturbances, chronic physical pain, depressive feelings, anxiety, and difficulty remembering or expressing thoughts (verbal amnesia). Furthermore, since trauma disrupts survivors' sense of safety and security, it is essential to offer a sense of reliability and collaboration during goal-setting. Lastly, understanding how trauma disempowers survivors affects the intervention process, and redirecting the conversation to collaborative discussion to restore their power and reaffirm their inner strength and resilience is necessary.

The use of social work skills learned during undergraduate and graduate programs, such as listening with care (paying attention with heart and mind to your client's stories), empathically understanding clients' primary concerns, open communication that allows clients to share information, and showing an understanding of cultural restrictions or limitations faced by the survivor, is a crucial part of providing TICBI. It is a collaborative process enabling clients to express their values and opinions about their lives. These social work value-based skills help social workers gain their clients' trust and increase motivation to participate in this healing process. Drawing on this established relationship, social workers move the conversation forward, creating a space for sharing feelings about the issues and the planned changes during each session. Providing TICBI has challenges, including occasional resistance from clients who revert to their adaptive survival responses for safety. For new graduates or early career social workers, the sudden resistance may be

N. Bokore, *Trauma-Informed, Culturally Based Intervention*,
https://doi.org/10.1007/978-3-032-06835-4_9

alarming, and some might even interpret it as personal dislike, blaming the client and labeling them as resistant, which is a term often given to clients from traumatized and marginalized communities.

Regardless of their career stage, social workers providing TICBI should be prepared for occasional resistance when working with survivors, irrespective of their mindset. Resistance is widespread when attempting to implement planned changes for survivors of gender-based violence, sexuality issues, and mental illness (a taboo subject in some cultures). It is crucial to shift the focus of the intervention to information sharing or psychoeducation to overcome this challenge and restore connection and collaboration. Utilize culturally respectful and evidence-based language to deliver meaningful information.

Even though your client's past trauma occasionally triggers setbacks in this phase, use your intervention skills of empathetic understanding and attuned listening to continue the work toward change. In time, TICBI can be expanded by introducing new and beneficial programs to your client, such as joining a group for additional support to gain from group knowledge. Since safety is important to survivors, be cautious about introducing new programs or overwhelming your clients with information they are not ready to process or accept. During the initial phases of the intervention, rely on your training and expertise to work collaboratively to provide TICBI and achieve the agreed-upon goals.

2 Paying Attention to the Layers of Sharing Information During TICBI

Motivational interviewing is one of the skills used in TICBI to foster change through skill-building and enhancing motivation to avoid misjudging client resistance. As discussed in Chap. 6, culture encompasses shared beliefs, values, traditions, and expectations, which influence the amount of information shared. When providing TICBI, the extent of shared information depends on this cultural membership and the established relationship between the social worker/therapist and the client.

When you think your client withholds information, remember that many communal cultures function with **three layers of information sharing**. The inner layer is confidential family information reserved for the in-group, the family, and the middle layer of confidential information with trusted friends. Moreover, the outer layer is reserved for outsiders with whom confidential family information is rarely, if ever, shared. As part of culturally based adaptation for family or group safety, these layers are purposely compartmentalized for increased protection from emotional pain or stigma. Receiving only outer-layer information during assessment and intervention from those with a survival mindset is common. The layers are further divided along gender lines. For example, when working with a male survivor, he may limit the information he shares based on gender, cultural rules, or taboo subjects.

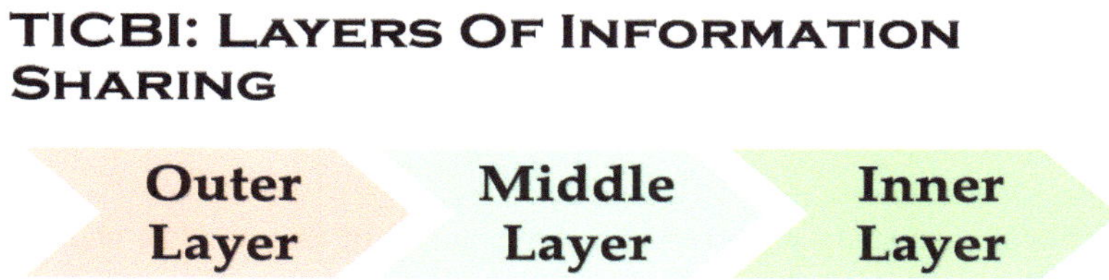

Fig. 9.1 Layers of information sharing

Another barrier to information sharing that might be mistaken as resistance is the presence of post-traumatic stress disorder (PTSD), which is a set of symptoms that are designed to protect (sympathetic response) because of memory loss or verbal amnesia. Understanding how a survivor's brain functions with limited memory is key to providing effective TICBI in these cases. Remember, memory issues create challenges for survivors, preventing them from adequately expressing their reaction to traumatic experiences. It is one of the challenges I faced after I arrived in Canada; it kept me emotionally isolated for over a decade until I finally learned about the benefits of therapy in my Bachelor of Social Work and graduate work (MSW and PhD).

In addition to cultural expectations, living with an active neuroception of danger slows down memory, increases fear, and prevents information sharing. Even after trust is established with social workers, sharing stories of gender-based violence is challenging because of the stigma attached to it. When providing TICBI, understanding the underlying issues behind information sharing, as illustrated in the process shown in Fig. 9.1, allows the social worker to work with their clients at their own pace without rushing, demanding excessive work, or misinterpreting the resistance. Doing so will break the relationship.

The TICBI approach involves understanding your client's thought process, actions, and emotions. Although some maladaptive behaviors affect their relationships, they are part of the coping strategies they developed to survive. You can help your client generate more effective behavior or responses to situations by teaching healthy communication skills and reducing stress and anxiety to improve their interpersonal relationships or social connections. The case studies chosen for this chapter are composites of refugees I worked with or met during one of my research projects.

3 Case Studies

3.1 Case Study #1: Survival Mindset—Ali

This fictional case study was created as an example of this mindset and does not apply to all male survivors from Syria arriving alone in a resettlement country. Ali, a single Syrian man, arrived in Ottawa, Canada, from a refugee camp in Lebanon in 2018. His medical doctor recommended psychotherapy. According to his doctor's referral notes, Ali is sleep-deprived due to nightmares and disturbing memories. During intake, Ali stated that lately he has started to isolate himself from people in his community to avoid recalling his past, and he no longer attends community events or prayer sessions at the mosque. He spoke negatively about himself using self-defeating narratives: "I am losing my mind, feeling hopeless all the time, and prefer to stay away from my community. I prefer not to discuss what happened or what is happening in my life. It affects my mood, making me sad and, at the same time, anxious."

3.1.1 TICBI for Ali

Review the initial assessment you did based on Chap. 8, exploring Ali's historical background and his journey in stages of forced migration (Chap. 2). Ali's official language is Arabic, which is spoken in most Middle Eastern countries with dialectic variations. In Ali's culture, having a large family is common, and not having any family members in Canada may contribute to Ali's distress and loneliness. Another concern is that Ali is currently single, approaching middle age, lacks financial stability, and is dealing with trauma-induced health issues.

3.1.2 Ali's Initial Assessment

Using steps outlined in Chaps. 7 and 8, the initial phase of TICBI for Ali's assessment begins with a bio-psychosocial, structural, and cultural evaluation, where the biological issues identified as trauma response and his stress level are good starting points for the initial talk therapy conversation. Ali's pressing problems are being distressed most of the time with intense physical and emotional reactions, creating self-isolation. In this and the following session, use attuned listening to identify the causes and symptoms of his distress, including PTSD-related responses and how significantly emotional dysregulation disrupts his daily life.

For some trauma survivors with a survival mindset, stress-related sensations become familiar, and they continue to struggle with the impact on their health of the consistent production of stress hormones (hypothalamic-pituitary-adrenal axis activation and increased cortisol production). For survivors with a survival mindset, these sensations are often mistakenly interpreted as a physical illness, causing frequent health service visits. Note the number of Ali's health-related complaints and doctor visits. If these visits or other maladaptive coping mechanisms are present, follow up with a conversation about coping skills as part of the intervention.

I have worked with survivors with a survival mindset who struggled with PTSD and sleep disturbance. Heightened sympathetic activation (a high arousal state due to the activated limbic region) is expected. Paying attention to Ali's survival narratives, including his daily life and relationships, is crucial. Working with survivors, it is essential always to provide a space for the relationship and ensure the quality of the intervention. Safety allows co-regulation and motivational interviewing about motivation, empowerment, and opportunities.

3.1.3 Ali's Intervention Goals

Short-Term Goals: Ali's intervention goals are developed by paying attention to his emotions and social and economic hardships. Avoid overwhelming or discouraging him from trying each treatment activity as you move on with the talk therapy. His short-term goals start with building trust, respectful communication, and prioritizing his goals. The purpose is to minimize triggers so that you can discuss fear-intensifying topics and engage in activities that include information sharing to help him regain a sense of his emotions and reactions. For survivors with a survival mindset, TICBI begins with a relationship and trust building with flexibility, renegotiating activities, or changing goals according to your client's needs, including long-term plans for changing their life and environment.

Long-Term Goals: Collaboratively creating long-term goals for change requires understanding Ali's communal culture and the impact of his adaptive social isolation on his social brain. When designing long-term goals, continue sharing information, for example, explaining how isolation is a slow death for him as someone from a communal culture. This is because of the decreasing white matter in his brain areas that regulate critical thinking and emotional control. This is because trauma is linked to the connectivity process between the amygdala and frontal lobes, creating difficulty assessing behavioral responses and the impact of maladaptive actions such as self-isolation. After all, humans are social beings, and their brains thrive on contact and connection (Cozolino, 2010, p. 177).

It is essential to remember that, regardless of their religious beliefs, spirituality, or cultural perspectives, humans are also driven by values. Therefore, considering or setting long-term goals for Ali depends on his motivation, which is linked to his core values. With this understanding, discussing Ali's core values and what matters to him now, as shown in Table 9.1, is essential. When reviewing his values and commitments to change, discuss potential challenges caused by his stress and ways to overcome them. Despite these anticipated challenges, such discussions aim to empower him and increase his motivation for change (Gordon & Borushok, 2017). Consequently, Ali's long-term goals should be specific, focusing on goals that are measurable, achievable, realistic, and timely (SMART).

Table 9.1 SMART goals for Ali

Specific	If Ali's goal is to have a good night's sleep every night, ask him how he will achieve this goal. Make sure the steps include specific evaluation tools to track achievement. For example, he might say I will do the following activities to evaluate my sleep pattern for X number of hours each night. Or he might say that to increase social connection, I will go to the mosque or community center X number of times per week
Measurable	Following the evaluation measure you set up, track the number of hours he slept each night and the quality of his sleep. Looking at the activities log for holistic and healthy changes, put a value on each activity and its stress reduction level based on a scale (1–5) of the relaxation methods used
Achievable	Evaluate each activity for change, including available support systems that have helped and can be used in the future. Pay attention to Ali's triggers as you create each goal and make it achievable for long-term change. Provide support along the way to avoid failure and decreased motivation
Realistic	Ask Ali how important social connection is to him based on his values and his commitment to change and improved well-being. Create intervention plans for Ali while paying attention to triggers and setbacks. Provide slow, step-by-step, simple plans or activities to avoid failure, decrease his hypervigilance, improve his concentration, and motivate him to work on change
Timely	Create possible end dates for each goal or activity, tracking the progress. Discuss setbacks along the way and methods used to overcome them. For Ali, the success of reaching goals is determined by how you collaboratively plan each agreed-upon goal with negotiable flexibility to change the goal and avoid feelings of failure and the activation of the dorsal vagal system

Review each short- and long-term goal, provide positive feedback when Ali accomplishes the required step toward making the change, and discuss the benefits of the shift to fuel Ali's motivation to change. Provide information (psychoeducation) as needed to help Ali understand the healing process in each step and the work ahead in making the change. Gentle and respectful motivational interviewing is helpful, considering his past experiences and his current emotional vulnerability. Motivational questions like the following are helpful for Ali's talk therapy activities.

- Can you tell me about your sleep patterns?
- You mentioned activities that sometimes help you sleep. Can you tell me about the specific activities and related thoughts or feelings associated with those activities?

At this point, theoretical integration will help you think critically, analyze Ali's trauma reaction, and plan possible interventions.

3.1.4 TICBI and Theoretical Integration for Ali

Multiple theories are used to inform TICBI services required for each mindset. For Ali, the psychodynamic theory developed by Freud can be used to explore Ali's unconscious responses to past traumas and understand how they impact his current behavior and relationships (e.g., Ali's self-isolation). Narrative therapy could help Ali weave together his unconscious and conscious reactions, giving him meaningful insight into his situation and the need to move toward change. Narrative therapy would allow Ali to evaluate his survival responses by externalizing what happened to him during forced migration rather than internalizing it. Narrative theory would allow Ali to use language and words of resiliency, survival, and motivation to continue living. Since dealing with cultural and gender-based learning is adding to Ali's stress level, opening a space for a discussion about his thoughts on cultural or religious/spiritual values and motivation to uphold them is needed.

For Ali, narrative theory can be used to reshape his thoughts about these issues and to assess his conscious and unconscious responses. It enables him to engage in a dialogue that relaxes him and helps develop psychological flexibility to examine his past, accepting painful memories and emotional wounds from previous experiences so they do not influence his future. Incorporating intervention strategies that include psychoeducation for self-understanding can be useful tools for survivors like Ali to foster change and growth through psychological flexibility and move away from emotions that create stagnation and impact well-being (Gordon & Borushok, 2017).

3.1.5 Psychoeducation for Ali

Providing Ali with TICBI services involves sharing helpful information about his past and present adaptive responses, as well as the increase in stress-related chemicals in his body, such as glucocorticoids produced by the adrenal glands, which play a role in re-experiencing traumatic events and the use of avoidance behaviors typical for survivors like Ali (PTSD). His intervention should include a discussion about the importance of these chemicals in surviving danger, as they activate the sympathetic response and prepare the body for action when danger is detected. However, the need to reduce the threat has passed. This will help him understand the functions of these chemicals and the impact their chronic production has on health, such as high blood pressure (norepinephrine), physical responses affecting body movement, lack of motivation, distorted perception of reality, inability to experience joy and pleasure (dopamine), and sleep issues (serotonin). Similarly, learning how to heal himself by balancing emotions through deep and mindful activities to reduce the chronic production of these chemicals is essential to correcting internal imbalances that affect Ali's homeostatic regulation and sense of safety (Cozolino, 2010, p. 265). The goal is to widen Ali's window of tolerance—the ideal state of emotional arousal where individuals can manage their daily lives effectively.

At this point, it is also essential to include information about the damage inflicted on his body by the extreme overload of the sympathetic system (active and ready for action) and the benefits of parasympathetic activation to increase calmness and motivation for change. This includes sharing information about managing triggers and adopting a flexible, calm mind and resilient outlook, allowing him to broaden his window of tolerance (Siegel, 2010). For Ali, this would improve his motivation to work on activities that will help him learn to tolerate triggers and intense feelings and practice homeostatic regulation to keep his body/brain sensations in balance. Providing information along with activities designed to reduce Ali's chronic stress (i.e., reduce the production of hormones by the active adrenal gland) and his extreme hypervigilance (self-isolation), paranoia, and perceptual distortions (fear of interaction with others in the community) will increase his self-understanding.

At this point in TICBI, psychoeducation for Ali would be increasing his awareness of his reactions and healing process to widen his window of tolerance (decreased hyperarousal and hypoarousal) which will help him to manage his lack of sleep, fear-related thoughts, and avoidance behaviors that decrease his curiosity about life (Ogden, 2021; Siegel, 2010; van der Kolk, 2014). However, it is essential to recognize when working with Ali that TICBI might be challenging, and he might be fearful and resist intervention. When that happens, it is critical to understand his resistance as part of his fear of change as a trauma survivor who is familiar with multiple and sudden life-changing experiences. Explaining resistance as part of Ali's adaptive hypervigilance helps him to self-understand, increase self-compassion, and refocus on the change process. In TICBI, client resistance is considered a trauma response that narrows their window of tolerance. You need to redirect the focus from the planned session to deal with the urgent need to help Ali regulate his emotions. For Ali, using a generalist approach when providing TICBI—seeing him from a strengths perspective and recognizing his survival skills, support systems, and innate ability to manage his life—helps start the healing process with information for self-understanding, reframing thought processes and activities to reduce stress, and increasing his homeostatic regulation and social network to widen his window of tolerance and promote healing.

3.2 *Case Study #2: The Therapeutic Mindset—Nimali*

Nimali, a young Sri Lankan married woman (she/her), came in to see a social worker. During intake, she described growing up in a multi-generational household and being consistently strong. At an early age, she also learned about modesty and obedience to her parents, loyalty to the family, and respect for gender roles in marriage. Nimali and her family left Sri Lanka during the civil war and resettled in Canada in early 2000. She experienced racism and microaggressions at school and in her neighborhood when she arrived in Canada, but eventually managed to adapt to the school system, thriving in her education. She completed her bachelor's degree in business and secured an excellent job. Nimali recently entered an arranged marriage and is four months pregnant with her first child. She came seeking services because of her recent experience with domestic violence. A few weeks into her pregnancy, her husband began abusing her and recently pushed her to the ground, knocking her backward onto a table and then onto the floor. She said she feels stunned, hurt, confused, and disoriented by the escalating violent behavior. When asked if she had told someone about the abuse, she said "no." She also said, "I do not know why I am having difficulty concentrating and why I cannot suppress memories of the past that began to pop into my head."

*Note that domestic violence is used as an example in this case study to demonstrate how triggers can affect survivors who are otherwise showing signs of healing and moving toward post-traumatic growth. This experience is not exclusive to South Asians and impacts many female survivors of war. The name in the case study and the country are fictional and created for this example.

3.2.1 TICBI for Nimali

Immigration and refugee resettlement are the two main factors driving population growth, shaping the new face of Canadian diversity. For example, Asian Canadians, including Sri Lankans, are the fastest-growing group in Canada after European Canadians, making up approximately 0.4% of the population as of 2021 (Statistics Canada, 2022). I developed this case study of a Sri Lankan woman from Southeast Asia because of the long-standing ethnic conflict between the Sinhalese majority and the Tamil minority in Sri Lanka, as well as the country's recent financial crisis, which caused mass population displacement from 1983 to 2009 and again in 2022. Unlike Ali's survival response, Nimali approaches her daily life with a sense of safety and a positive outlook rooted in her post-trauma resilience, which has helped her manage her emotional reactions. Shortly after her arrival, Nimali adopted a therapeutic mindset. This constructivist approach enabled her to adapt to her new environment by relying on her core beliefs about herself, her future, and the world around her. When providing TICBI services, it is important to recognize that clients with a therapeutic mindset often appear ready to communicate and share information. However, they also experience underlying emotions connected to past experiences, such as sensitivity to triggers, active sympathetic responses, and alertness to potential danger.

3.2.2 Nimali's Assessment

In addition to forced displacement and trauma, Nimali is experiencing intimate/domestic violence, taking her back to her survival-mindset space from her otherwise therapeutic mindset. Domestic violence, which includes diverse methods of abuse, including aggression and intimidation, is a widespread occurrence that takes place in all cultures, ethnicities, and nationalities. However, immigrant and refugee women, particularly survivors of wars, are especially vulnerable to the impact of domestic violence because of their past traumatic experiences. For some immigrants and refugee women from communal cultures, the challenges of gender-based restrictions become added barriers, prolonging their abuse. Review your intervention plan to understand Nimali's situation (Chap. 7).

- Conduct a bio-psychosocial, structural, and cultural assessment. Are there factors that stood out for you linking past vulnerabilities to the domestic violence she is experiencing?
- Explore the presenting issues, Nimali's personal strengths, resilience, and resources.
- Since Nimali was identified as having a therapeutic mindset during the assessment and exploration phase, explore her values and motivations to make a change.

As you review her assessment, explore issues linking her current situation to her past experiences. Since Nimali's case is complex, reflect on your knowledge about trauma, post-traumatic growth, domestic violence, and refugees and set goals for TICBI. Setting goals is challenging for Nimali, who is now dealing with triggers from her past and stress about the safety of her unborn child and the cultural stigma about a failed marriage. These issues may affect her internally, awakening her danger alerts and sensations, creating additional stress, and impacting her cognitive processes. Listen to her narratives and thought processes as she collaborates with you on setting her short-term and long-term goals.

3.2.3 Nimali's Goals

Short-Term Goals: Based on Chaps. 7 and 8, short-term goals are needed to provide psychological first aid to activate Nimali's cognitive thinking. It requires sharing information about calming exercises, such as breathing, to decrease her stress and reduce the overwhelming emotions causing her disorientation and distress. Immediate resources are needed for her safety, shelter, income, and family support. After her emergency is clarified, the TICBI intervention will be moved to setting and reaching long-term goals, which include information on supportive systems with an increased awareness of her safety needs and the steps she needs to take to achieve her goals.

Long-Term Goals: In situations of domestic violence among refugee families from patriarchal communal cultures, where women are often perceived as the cause of the violence, it is crucial to review Nimali's history and culture for a respectful and informative conversation. Understanding her perspective on family issues influenced by religion, cultural taboos, and concerns about violence is essential. In this case, the long-term goal of TICBI would be to uncover the layers of cultural constructs normalized in the client's life and their effects on well-being. Some of these beliefs encompass a gendered and powerful concept that states the husband wields power in the household, implying that violence is accepted or necessary in some cases to resolve marital conflict. It is vital to deconstruct beliefs learned in the past but no longer functional in the present.

However, it is also necessary to be mindful of how domestic violence is hidden from those perceived as part of the outer layers of information sharing. Individuals experiencing domestic violence are often obliged by culture to remain silent. For those who have undergone forced migration, as Nimali has, the situation becomes more complicated because families depend on one another for safety. The fear of going against family and cultural norms can create intense and paralyzing anxiety, impacting Nimali's decision-making abilities. Therefore, when establishing long-term goals, addressing the client's fear-based thoughts, hesitation, resistance, and emotionally charged explanations is crucial, considering the following concepts.

- The success of the long-term plan relies on Nimali's motivation and what is important to her both now and in the future, including her family as a key part of her support system.
- Activities should focus on rebuilding Nimali's positive self-image, re-igniting her strength, and resetting her post-trauma growth with greater confidence, self-empathy, and self-belief.
- The long-term intervention also encompasses the practice of assertiveness and communication grounded in cultural respect, fostering an exchange of feelings, ideas, and futures with families and significant others.

Talk therapy has short- and long-term goals. These include accepting the current reality and the challenges ahead to minimize disappointment and prepare Nimali to modify her goals to achieve successful outcomes. Nimali's case is complex and has multiple variables. Therefore, the goals and intervention plans are prioritized based on necessity.

3.2.4 TICBI for Nimali

To achieve Nimali's long-term goals, continue to work with her, providing emotional support and co-regulation to move back to a therapeutic mindset. The shared information and activities should include knowledge to increase her ability to evaluate her unique situation and motivate her to change. I say unique because immigrant and refugee women's experiences are varied according to their culture, status of migration, and support systems. Therefore, awareness of Nimali's response to domestic violence is needed to identify her basic needs, core values, and reasons for seeking a change. Exploring those values and reasons is part of the TICBI process. For Nimali, self-understanding, including the benefits of reactivating her therapeutic mindset, will motivate her to continue practicing change-based activities and actions. Use focused and empathetic listening throughout the process, paying attention to the sources of her strength and the resilience she used in the past to navigate school and neighborhood challenges successfully.

One thing I have learned as a survivor of war, later a service provider, and now a researcher, is that women wear multiple hats and compartmentalize their trauma responses to protect their children and families. I have categorized women's adaptive trauma responses into four survival skills that can benefit Nimali and help her achieve her long-term goals. It begins with asking her to identify which category she would place herself in and designing the intervention accordingly. In TICBI, working on domestic violence cases is not about eliminating the violence—which takes time and extensive work with communities—but about empowering individual victims to use newly acquired knowledge and resources to overcome their past violent and traumatic experiences. Table 9.2 illustrates the four informative and adaptive skills that can be employed during the change talk discussion.

If Nimali identifies herself as a strong and resilient woman with adaptive skills, the TICBI intervention will focus on reawakening her resilient, strong self. Suggest activities focusing on her cheerful and adaptive responses, including relearning to tune in to her internal messages, assessing her current situation from various perspectives, and using her adaptive response and wisdom. This will help her find an anchor of safety and the adaptive self within her, allowing her to use her cognitive thinking to find solutions to save herself and her unborn child.

Change talk for Nimali is about self-understanding. For those with a therapeutic mindset, it must include focused listening to narratives of survival and identifying the characteristics defining each client based on the above-listed categories.

TICBI talk therapy for Nimali also includes listening to her values, motivation, and capacity to follow through with her short- and long-term goals. Cultural and religious teachings are at the forefront of these discussions. In Nimali's case, the conversation might lead to the Hindu description of suffering as an aspect of karma from the past that will be endured with the goal of spiritual liberation later (Hadad,

Table 9.2 Women's adaptive trauma responses

The traumatized woman	A traumatized woman is stuck in the sympathetic, or dorsal vagal, system, experiencing the multiple impacts of her experiences: emotional difficulties, fear, anxiety, depression, and a sleep disorder. For a woman who is also dealing with domestic violence, her past and present exacerbate her emotions, keeping her in the survival mindset. Working with a traumatized woman requires understanding her situation and providing educational resources to promote healing and help her navigate the complex challenges of her life. This includes learning how to sort out the sources of her emotional responses, self-regulation, and working toward change
The resilient woman	The resilient woman looks at life through possibilities. She uses her values, motivation, and opportunities to challenge life circumstances. She has a change mindset, and she is confident in her abilities. She uses her analytical self to assess her situation and solve her challenges. She has a confidence often noticed by others, who also become her support system, encouraging her to move forward with the change in her life circumstances, including domestic violence
The adaptive woman	The adaptive woman has similar characteristics to the resilient woman with an added positive outlook on life, and this helps her to recognize the need for change when confronted with life challenges (including handling domestic violence and making a decision about it) by acknowledging the need for external and internal support that will further her post-trauma growth
The warrior woman	The warrior woman has the character of a resilient, assertive, and adaptive woman. She effectively learns positive and adaptive responses to find solutions for her current situation. She uses protective factors such as her support systems or an optimistic outlook on life to stabilize her emotions. She manages her past traumatic pain with grace and acceptance, looking at her past physical and emotional wounds with self-compassion that allows her to move forward with confidence. Her creativity, connection, self-compassion, and social support move her toward overcoming challenges and healing

2013). A conversation about religion and spirituality with Nimali will be much easier than with women from other religions, as she is familiar with the practice. As Eastern religious practices have become increasingly common in Western culture, engaging in Eastern religious activities, such as yoga, Ayurveda, and meditation, as healing practices has become a regular part of everyday life in Western-based therapeutic settings.

Alongside self-reflective conversations and practices, it is important to address Nimali's experience of domestic violence and how it triggered what Bowlby (1951, 1973) called an attachment injury, or breach of trust. She has significant knowledge that can improve her understanding of her emotional responses to her situation. For adults in similar situations, early experiences often act as trigger points that activate their fear responses, shaping their reactions to current events and causing the shutdown of their cognitive responses as their bodies and brains react to what is happening (Bowlby, 1951; Ogden & Fisher, 2015).

When traumatic events occur for survivors of wars like Nimali, they trigger her past autonomic nervous system (ANS) responses, leading to emotional dysregulation and maladaptive behavior. Activities that decrease dysregulation (see Chap. 10) include self-reflection, grounding exercises to stimulate the growth of new neurons, and cognitive responses that reshape thought processes to facilitate change. Self-reflection in social work practice enhances Nimali's awareness of the past while focusing on the present moment, encouraging her to look toward the future and strive for change. For example, self-reflection re-ignites Nimali's resilient and adaptive self by focusing on using daily logs (Table 9.3) to record her life and reactions to triggers.

These daily logs will also assist Nimali in focusing on her triggers as she identifies their sources and works toward changing her life. Sharing encouraging information about the connection between Nimali's present and past experiences, gender

Table 9.3 Template for a daily, self-reflective log

Activities	Responses
Time and date	
As you experienced triggering emotions, what did you notice happening within yourself and your internal sensations? Do you feel anxious, angry, or frozen when confronted with stressful situations? If so, are these emotions or sensations familiar, or have they developed recently?	
Reflecting on your relationship with your partner, what feelings, emotions, and sensations are you becoming aware of or experiencing? Are these connected to current experiences with your partners or past stressful situations?	
Reflecting on your past traumatic experiences and associated emotions or physical sensations, what activities did you engage in to overcome them?	
As you reflect on your adaptive responses, including those influenced by cultural or religious teachings, what steps do you need to take to change the negative responses that are affecting your well-being?	
Suppose you bring out your warrior woman (see Table 9.2) that helped you survive the war and adjust to resettlement challenges. What steps do you need to take to revive it, and how will it assist you in making a change in your current situation?	

roles, and violence is beneficial for her self-understanding (Hecker et al., 2015; Johnston, 2011). Psychoeducation in domestic violence cases, which involves defining the emotional impact of violence, is an evidence-based conversation that is culturally appropriate, using scientific knowledge and shared through activities designed to expand Nimali's window of tolerance. The collaborative process of information sharing increases Nimali's sense of safety and self-regulation, reduces her feelings of helplessness, rekindles her resilient self, and, in the long term, supports her in using her strength and knowledge gained through the TICBI process to make positive changes.

3.2.5 TICBI and Theoretical Integration for Nimali

Experiencing interpersonal violence from a partner can heighten the risk of triggering PTSD symptoms or the emergence of other mental health disorders in survivors of past traumas. Therefore, creating a theoretically informed intervention for TICBI is essential to understand this risk. Addressing gender-based violence and intervention plans using feminist theories helps to explain the core schema of intimate violence and the power dynamics that precipitate coercive and controlling actions against women (Johnston, 2011). For immigrant or refugee women arriving from the Global South, the gendered and culturally learned patriarchal and religious norms are rooted in guilt, duty, and shame, where the blame on women is more pronounced. Consequently, in some of these cultures, cultural and religious values also defend male partners' emotionally and physically abusive behaviour, hiding it beneath cultural layers of information sharing (see Chap. 8) for issues considered taboo subjects that are not disclosed to outsiders, including counsellors or social workers.

For TICBI, using a feminist perspective allows social workers to increase awareness of power differences and the role that culturally and spiritually dictated social learning plays in families, particularly those experiencing intimate partner violence. Social learning theory explains how these learned behaviors are internalized at a young age and emphasizes the need to understand their impacts and consequences on families, especially those facing power differentials and intimate partner violence. Social learning theory, developed by psychologist Albert Bandura (1977), suggests that social learning occurs through observing others in the social environment, including parents. In cultures where women are seen as property, the acceptance of aggressive and power-driven behaviors in males is widespread, with no relevant consequences. Therefore, for Nimali, this theory assists in discussing gender roles in society, their impacts, and the risks of conditioned behavioral responses.

It helps to discuss the benefits and risks of culturally informed learning as part of patriarchal rules and structures, and how what is learned can be unlearned. Viewing domestic violence as an expression of power and the risks associated with that power at this particular time in her life will help Nimali examine the familiar social structures that contribute to her husband's behavior (Lawson, 2012). Similarly, evaluating domestic violence using both social learning and social exchange theory will assist Nimali in understanding her husband's violence and the structures that maintain it. When discussing the sources of gender-based violence, including domestic violence, it is crucial to focus on behavioral responses for self-understanding, healing, and recovery, providing a well-defined explanation and activities to mitigate fear or re-traumatize the client.

3.2.6 Psychoeducation for Self-understanding

In domestic violence cases where cultural restrictions limit the victim's actions, psychoeducation helps to increase awareness of the lethality of the violence. Providing information about physical and psychological safety helps survivors like Nimali expand their decision-making abilities with options that empower them, moving them away from culturally reinforced fight/flight (hyperarousal) or frozen (hypoarousal) states. As you create a safe therapeutic space in each session, widening Nimali's window of tolerance and activating her resilient adaptive reactions, also examine her perspectives on negotiating the risks of cultural learning in the presence of domestic violence. Explore how Nimali views her role in maintaining her family's honor or fulfilling her cultural duties related to marriage and family life. Does this desire or obligation supersede her individual needs at this time in her life?

In this conversation with Nimali, please pay attention to the cultural values that are important to her while respectfully raising awareness about the dangers of domestic violence during pregnancy. At this point in TICBI, it is essential to focus on her perspective regarding safety and how she articulates the threat of domestic violence. In cases like hers, where culture is central to family structure, the Rogerian client-centered approach involves acknowledging clients' expertise in their lives and respecting their choices. Therefore, make each interaction a collaborative effort, recognizing the dangers of domestic violence, acknowledging her inner resilience and strength, and outlining the external support systems available to her, including family, friends, and programs.

The goal of TICBI at this stage is to address Nimali's needs and discuss the complexity of her traumatic past and present. This conversation allows her to reflect on and reconsider her immediate needs. It encourages her to utilize the available resources while creating emotional balance (expanding her window of tolerance). Track her responses and engagement with the activities that helped her regain control and foster calm and reflection. Note how these activities moved her from a sympathetic (fight or flight) to a parasympathetic nervous system (relax and connect) response. Pay attention to how Nimali is now able to change her narrative as she further understands her emotions, values, motivations, and what she needs to do to cope with her compounded trauma.

It is a reflective process in TICBI designed to increase her understanding of the situation and her emotions and empower her to make changes. These changes will re-ignite her previously adaptive response (being at the therapeutic mindset), regulate her emotions, and promote post-traumatic growth, regaining her strong, resilient self.

3.2.7 Psychoeducation for Regaining Strength

Psychoeducation is an integral part of TICBI when working with survivors of wars and dealing with domestic violence. It helps them reduce triggers from their past, regain balance, and understand their current risk. It provides the knowledge and skills needed to survive and build healthy relationships. It facilitates learning how to stabilize emotions and regain empowerment, especially for those disempowered by cultural or social conditioning. It enables them to reclaim their sense of self and make informed decisions for change. Western and Eastern philosophies espouse various methods for regaining strength and stabilizing emotions or relearning self-regulation after experiencing dysregulation due to recent crises. Mindfulness activities help survivors remain in the present moment and focus on what they value most (Hadad, 2013). These activities are delivered in multiple formats, from social media to books to community training sessions, making them accessible to Nimali.

To decide to change her current circumstances, Nimali needs to learn and practice mindfulness to gain the emotional brake she needs to pause and regulate herself, giving her a chance to evaluate her situation and make a choice. The emotional brake increases the chance for prefrontal cortex activation, focused attention, and emotional stabilization (Schore, 2012; Siegel, 2010). For Nimali, this process also helps her engage her thinking brain, regulate her emotions, and take appropriate actions to protect herself and her baby. Adding mindfulness to her intervention refocuses her attention, moving her away from her overwhelming fear of domestic violence and encouraging her to pay attention to the present moment, where she needs to protect herself and her baby. In domestic violence cases, self-regulation means regaining self-control and managing implicit reactions to her present experiences.

The added benefit of mindfulness practice for Nimali is the expansion of her window of tolerance (Kabat-Zinn, 2005; Mitchell & Wupperman, 2023; Ogden & Fisher, 2015), utilizing her inner strength (identifying herself as a warrior woman) and applying skills gained in her TICBI sessions to self-regulate, self-reflect, and manage her emotions (Kabat-Zinn, 2005). This practice will also help her increase her hippocampal activity to bring her past adaptation of emotional balance and growth (therapeutic mindset).

Mindfulness activities as part of psychoeducation will help Nimali use breathing techniques (see Chap. 10) and prenatal stretching activities to learn how to calm herself and activate her vagal brake and parasympathetic response, motivating her to make changes (Porges, 2007, 2017). TICBI for survivors who experience domestic violence and have underlying trauma is about regaining strength, clear thinking, self-compassion, and self-understanding, focusing on values and what is needed in the present moment, as well as what needs to change in the future to decrease stress. For those with a therapeutic mindset like Nimali, it is also about reflecting on past motivations for moving toward growth and accessing resources on the path to recovery.

3.3 *Case Study #3: Active, Engaged Mindset—Sarah*

Sarah is a woman in her 50s, born and raised in Ethiopia, who recently lost her long-time partner to suicide. She arrived in Canada through a Canadian humanitarian program after escaping political violence, ethnic cleansing, and food shortages. Despite her experience of early childhood trauma and recent displacement, Sarah was doing well until her husband suddenly and unexpectedly took his own life. When she arrived, she connected with Toronto's Ethiopian and Orthodox communities to secure housing, employment, and social support. Sarah and her partner benefited from their strong culture and opportunities in Canada and have done well in the past two decades, rebuilding their careers until her partner's passing. She is concerned about her recent emotions, which she describes as "down in the dumps." She reported having difficulty organizing the burial and an inability to eat and sleep due to nightmares about her husband's suicide.

*Please note this is a hypothetical case study created as an example and does not suggest that suicidal ideation is common among Ethiopian refugees. Suicidal ideation, unfortunately, occurs in families of war survivors, who deal with the complexities of trauma and triggers.

3.3.1 Sarah's Initial Assessment

Providing TICBI for survivors like Sarah involves exploring her response to the recent crisis, including the presence of complicated grief or mental health issues. Evaluating her emotional and psychological reactions due to the loss of her partner by suicide, the level of support around her, and the biological factors that contributed to her internal strength, which motivated her in the past and allowed her to function with an active, engaged mindset, is a crucial aspect of the intervention.

When working on cases like Sarah's, social workers often utilize the grief/loss understanding of Western cultures, known as the Elisabeth Kübler-Ross (1997) model. This model is based on stages of loss categorized into five responses: denial, anger, bargaining, depression, and acceptance. However, this list and process vary across cultures. Eastern and Western cultures differ in their grieving approaches due to religious beliefs, social learning, and perspectives on life and death. Exploring these distinct approaches is important for identifying the grieving process. For example, while Eastern cultures embrace collective mourning and follow rituals (maintaining a connection with the deceased), Western cultures often experience individualistic grieving, supported by the process outlined in Elisabeth Kübler-Ross (1997) model.

Sarah is showing distinct cultural grief responses. It is also evident how her recent loss of her long-time partner, who supported her during past traumatic and life-challenging experiences, impacts her. She carries extreme guilt for not noticing or recognizing the signs of his suicidal ideation and depression. In Ethiopian culture, where norms and values are informed by the two main religions, Islam and

Christianity, both belief systems stigmatize suicide or self-exit, creating taboo conditions that keep those with suicidal ideation and their families silent and without support. As a result, Sarah may initially limit the information-sharing process in her initial assessment and intervention. The stress and fear of keeping the secret of the method of her partner's passing left her without adequate community support. Traditional mourning for Ethiopians varies based on ethnic and religious groups. However, all groups share the process of lengthy mourning periods and extensive rituals designed to remember and connect with those who passed and support the grieving families left behind.

While keeping the impacts of these challenges in mind, paying attention to her internal strength will help in working toward a goal that will prevent her from developing complicated and prolonged grief, re-igniting her post-traumatic growth (Maciejewski et al., 2016). Collaboratively with Sarah, create goals (short- and long-term) based on her warrior and adaptive woman behavior. To make it simpler and more motivating, ensure these goals are specific, measurable, achievable, relevant/realistic, and timely to assist her with mourning, grief, and coming to terms with her loss.

3.3.2 Sarah's Goals

Short-Term Goals: Remember, Sarah's sudden loss created a new crisis in her daily life, changing her social status from a married woman to a widow. Therefore, include psychoeducation about grief and loss in the initial sessions to enhance her self-awareness of her current grief responses, previous survival energy, her post-traumatic growth, and how to overcome her grief-based setbacks. When setting short-term goals with Sarah, discuss accepting loss as part of the human experience, including what can be learned from religious and cultural teachings or mindfulness practices. The short-term goals could include

- Accepting her loss and learning to live without her partner, she adapted to living with the memories of her now-deceased husband.
- Accepting death as one of life's inevitable crises is common to all living beings.
- Processing the pain of her loss and the negative emotions that come with the grief.
- Use mindfulness activities to shift her focus from her loss to a celebration of her husband's life. This will ensure a vagal brake and activation of her parasympathetic nervous system to decrease her physical and mental stress and promote relaxation, opening her thoughts to the possibilities of new and positive life experiences on her own.

As you establish these short-term goals collaboratively, use empathy, focus, and compassion when listening to her story, and provide supportive energy (co-regulation) to help Sarah share her story comfortably and plan a life after her loss. At this point, TICBI is about providing Sarah a chance to collaborate for change by ensuring the level of information sharing is at the middle layer (see Fig. 9.1). Since grief functions in stages that are sometimes cyclical (based on the stress level and triggers), validate Sarah's emotions, acknowledging her current physiological struggles. Focus on positive cultural, religious, and spiritual teachings to help Sarah see new possibilities while celebrating her partner's life and his contribution to her life. Having these conversations during the initial phases of TICBI will allow Sarah to think about coping activities that could be included in her long-term goals, such as joining a group counseling session for additional support and increasing her chances to develop "psychological flexibility" and "acceptance" of her loss (Gordon & Borushok, 2017, pp. 47–48).

Long-Term Goals: Part of achieving her goals is helping Sarah accept the tragedy and loss, reframing her thoughts and remembering the positive aspects of her married life to re-ignite her resiliency and her spirit of survival. Sharing positive memories will help her learn to live without her partner, modulate her emotional state (ventral vagal circuit), and adapt to her new life.

Learning about emotional regulation also allows her to function again within her window of tolerance as she remembers life with her partner without being overwhelmed by grief. It is a re-adjustment of her emotional dysregulation and fear of facing the future alone, anger about her partner's last action of taking his own life, anxiety (hyperarousal) about it, and feeling empty and hopeless (hypoarousal). The long-term goal of surviving this loss includes activities such as:

- Mindful breathing to help her with her grief and stress about facing the future alone. Mindfulness is an acceptable practice recommended by most religions, with varied instructions on how to do it. In most cultures, it includes breathing, where taking deep and slow breaths into and out of the abdomen is used to center emotions. It increases focus and brings calm to grief-related emotions. (Caution: This might be an anxiety-inducing practice for some clients with breathing issues. Therefore, start slowly until your client finds their comfort level.
- Increasing community-related social activities to encourage Sarah to reconnect and receive support.
- Physical activities, such as rejoining the gym she was part of before her partner's passing. This will help her reduce stress, provide a distraction, release endorphins, improve self-esteem, become part of self-care, and help her return to a regular sleep cycle.
- Continuing with the change talk to help Sarah work on her distorted thinking about losing a partner, the future (life without him is no life), and self-blame for not recognizing his suicidal ideations.
- Connecting her with others in a similar situation (group counseling), allowing her to learn from others by discussing her loss and expressing the emotions that may arise from time to time in a supportive environment.

Building these long-term skills for surviving the loss of a partner will help Sarah overcome the current traumatic change in her life and transition to surviving by engaging in future-oriented talk. Positive and healing talk will also help Sarah reactivate her parasympathetic nervous system and use her previously favored circuit (ventral vagal), moving her away from the survival mode (sympathetic nervous system) she is currently in. The successful transition can be achieved as part of the long-term plan through psychoeducation to encourage her to reconnect with her community as well as her mind and body.

3.3.3 Theories Informing Sarah's Intervention

The loss of a loved one is a challenging issue to discuss during TICBI. Past wounds resurface, re-igniting the pain of previous losses, triggering PTSD and depression. The Kübler-Ross (1969) model can be used for TICBI to explain Sarah's complicated grief and current psychic response, which includes sadness, fear, numbness, anger, and guilt. For Sarah, the loss triggered sudden memories of losses due to conflict and during the initial stages of forced migration. The Kübler-Ross (1969) theory opens the door to discuss the impact of her sudden loss. As mentioned above, even though grief is a universal experience after a loss, people handle it differently based on culture and religious beliefs.

In Sarah's case, using a variety of theories to inform TICBI is helpful when integrated with cultural practices that promote a holistic and growth-promoting model based on her strengths. For Sarah, this would involve a discussion about the grief process, including the acceptance of her loss and a focus on personal growth. This conversation will assist her in processing her sadness, attending to the emotions related to the loss, and refocusing her thoughts by concentrating on the positive aspects of her relationship with her partner. Such a discussion will enable her to overcome emotional disequilibrium and crisis (depression, anxiety), while focusing on relearning and re-igniting her post-traumatic growth. Using grief theory based on cultural communication allows for exploring and negotiating its application in mourning and the cultural expression of grief. For Sarah, employing the Kübler-Ross and the post-traumatic growth models will help her refocus on survival and adaptation while participating in activities that create balance, self-reflection, and recovery.

3.3.4 TICBI for Sarah

For Sarah, the discussion about the Kübler-Ross model and refocusing on what helped her develop post-traumatic growth provides an opportunity to reflect on her transition from current loss to empowered self. This conversation involves considering her past positive outlook on life, her resilience, and her post-traumatic growth, which will determine the outcome of TICBI. Begin the conversation by discussing the process. In social work clinical practice, bereavement and grief are often treated as the same. However, while bereavement defines the experience of losing an important person, as Sarah did, grief defines her reaction to that loss, which in TICBI becomes the primary focus of the intervention. Therefore, it is essential to keep in mind Sarah's grief and her reaction to the loss. Thus, for Sarah, TICBI will include

- Discuss her lifelong priorities and support systems, such as strong religious beliefs, culture, and community, which have helped her manage or overcome grief-stricken emotions in the past.
- A conversation about engaging in culturally acceptable mindful practices, such as deep breathing and relaxation techniques, to help her reduce the stress, fear, and sadness stemming from her past and present losses.
- A conversation about participating in community activities, such as group readings of religious scriptures and Orthodox Christian teachings, was something she used in the past to help her balance her emotions (acceptance and hope).

For Sarah, the conversation will also discuss the connection between grief and body/brain reactions, initiating changes in her autonomic response. Grief causes neurological changes that include the release of stress and sadness-induced neurochemicals and hormone production, causing physical exhaustion, loss of appetite, sleep disturbances, and anxiety. In TICBI, psychoeducation explains the fascinating human brain and its extraordinary ability to sustain life by taking new sensory information (such as information related to recent loss) and sorting it out according to its importance and previously stored knowledge (van der Kolk, 2014). For Sarah, the conversation about stored knowledge includes remembering what motivated her in the past to work on her survival, adaptation, and growth.

3.3.5 Psychoeducation for Sarah

For Sarah, the discussion about brain-based responses to grief involves a conversation about structural changes, especially in the underactive or overactive areas of the brain. For example, the prefrontal cortex (reasoning and decision-making) becomes underactive during grief, creating an overwhelming sensation of sadness that impedes concentration. In this process, the anterior cingulate becomes underactive, affecting the integration of emotional information and modulating the function of the autonomic nervous system. Sarah may be triggered by her past losses and experiencing the overactivation of her amygdala (fear circuit); the life-changing loss she experienced may increase her fear and panic about the future. Thus, it is essential to activate Sarah's socially connected brain by including group activities to help her rewire her brain, returning her to her pre-recent loss state and active, engaged mindset. This reactivation process includes exploring and using culturally acceptable healing practices (see Chap. 10 for details), such as breathing exercises, which can help Sarah focus on calming her emotions.

3.3.6 Culturally Informed Healing Practices Exercises for Sarah

Culturally informed healing practices for coping with the stress of loss are part of self-care used in various cultures and religious practices. For Sarah, these culturally informed healing practices will help her focus on the present moment, reduce the intensity of her emotions, and refocus her attention on what is important to her at the present moment. For example

- Focusing on the here and now during breathing exercises will help her manage her intense emotions and visualize her future while keeping her special memories in the box where she keeps all the good memories from her past for comfort and acceptance.
- Using the Ethiopian method of sharing stories over three cups of strong Ethiopian coffee allows survivors to use the oral tradition of grief sharing, celebrating life, and reflecting.
- Writing thoughts in a memory book while acknowledging grief as a process (stages of grief and loss) and its emotionally challenging effects.
- **Participating in group activities like mindfulness walks in the park helps** Sarah feel connected to the comfort of nature and the present moment. Walking through neighborhood parks, smelling the flowers, observing nearby water if available, and listening to the sounds of birds will help Sarah ground her emotions and engage with her senses and the world around her.
- **Other physically engaging, culturally informed healing activities that** provide Sarah with a vagal brake to calm her emotions and establish balance (see Chap. 10) by reducing her fears of loneliness, loss, depression, and the stress of the unknown.
- Introducing Sarah to connections with nature, religion, culture, and support systems helps her function within her window of tolerance when she is temporarily unable to do so. The use of culturally informed healing practices helps Sarah expand her window of tolerance by reducing her anxiety and fear of the future without her partner. The activities also provide Sarah with a vagal brake, allowing her to re-ignite her growth process.

These activities will be acceptable to Sarah, as breathing through intense emotions is common in many cultural and religious traditions. The process will help Sarah reactivate her thinking (prefrontal cortex) and memory (hippocampus) centers, including the neuroception of safety, and shift her from a sympathetic response to a ventral vagal state. This becomes the foundation for increasing her ability to regulate her emotions, regain resilient behavior, and improve her flexibility in navigating her autonomic hierarchy and responses.

4 Reflection and Evaluation of TICBI

Now that you have used TICBI in the above three case studies, it is essential to include an evaluation process that begins with a self-evaluation of the skills you used and the benefits of their use. Multiple techniques are available to evaluate your intervention skills, including online and paper-based evaluation systems. An online system helps to track your skills and the benefits of the intervention. It is also easier to convert the evaluation outcome to visual scores and tables or create graphs and charts. To generate a paper evaluation form, ask yourself the following reflective questions and assess yourself using a rating scale of 1–4.

1. is a rating that shows low skill in this area and highlights the need for improvement.
2. shows an average rating and needs further training or reading for improvement.
3. an indication of average knowledge with room for improvement.
4. more than adequate knowledge in this skill.
5. exceptional skills.

- Did I stay present during the intervention process and track my client's moment-by-moment experiences?
- Did I increase connections and use focused, empathic, and attuned listening to identify my client's needs?
- Did I communicate clearly with my client to explain the process and engage in here-and-now change talk?
- Did I engage my client in an open conversation, address their concerns, and thank them for their valuable feedback?
- Did I provide psychoeducation and other necessary information to support my client?
- What skills have I used when working with clients with each mindset? Evaluate each skill and benefit based on client feedback on a scale of 1–10. One is the least benefit for your client, and 10 is the most significant benefit for your client.

I often use this type of self- or group assessment rating scale for in-class activities in clinical skills.

5 Termination/Ending TICBI Sessions

TICBI sessions end with a review of the benefits of the treatment plan and an assessment of whether further support is needed, such as referrals or ongoing assistance from a specialist (e.g., mental or physical health professionals). It is important to start this conversation early to help clients prepare for the emotional finality. Since many clients have experienced multiple losses, this preparation helps them to celebrate their progress and understand the significance of concluding the sessions. It also encourages clients to apply the skills and insights gained to future challenges. Ending a relationship is difficult, but a successful conclusion is ideal. However, there are times when ending or terminating TICBI sessions becomes challenging, especially if the client wishes to discontinue or drops out of the intervention. In such cases, following up with the client is crucial. Although difficult, particularly with survivors who have experienced power imbalances, follow-up calls are necessary. It is important to remember that some survivors develop a tendency to please authority figures as a survival strategy, which often masks the actual reason for dropout or termination. Therefore, TICBI providers must pay close attention to their clients' responses during calls and be attentive to their emotions. With sensitivity and understanding, it is possible to obtain feedback on the service provided, which benefits the program and also helps evaluate your skills, methods, and relationship-building process. This final feedback in TICBI allows social workers to improve their practice and make positive impacts on survivors' lives. Evaluating TICBI sessions supports social workers in updating their skills with the latest research, knowledge, and professional development. As you practice TICBI, use case studies as a resource, and review what you have learned from this chapter by answering the discussion questions.

Definitions

Activated limbic region: The brain system, or area, responsible for regulating emotions.

Co-regulation: When the nervous systems of two individuals interact to support one another and emotionally connect through the mutual regulation of their physiological states.

Dopamine is a reward chemical known as the "feel-good" hormone, providing a sense of pleasure and motivation.

Endorphins are chemicals (hormones) released during pain or stress or while doing pleasurable activities, including exercising, getting a massage, eating, and having sexual intercourse. Endorphins help relieve pain, reduce stress, and improve well-being.

Glucocorticoids are anti-inflammatory hormones found in all tissues that help regulate muscle metabolism in the liver and bones. These hormones, produced by the adrenal glands (which are part of the hypothalamic-pituitary-adrenal (HPA) axis), play a complex role in the body's response to trauma. Part of their intricate functions during recovery includes contributing to traumatic experiences, which can lead to psychological conditions like PTSD, as well as supporting recovery, processing, adaptation, and healing.

Norepinephrine: This hormone, noradrenaline, transmits nerve signals from nerve endings to other nerve cells, muscle cells, and gland cells.

Serotonin *is t*he chemical (hormone) that carries messages between the body's nerve cells and the brain. It plays a key role in mood, sleep, digestion, wound healing, and bone health.

Discussion Questions

1. This chapter explores layers of culturally informed communication. In brief, explain the significance of this form of communication as well as the challenges of collaborating with diverse populations that parallel the cases discussed in this chapter.
2. Why should you pay attention to the culturally structured layers of sharing information when working with survivors of prolonged wars and forced migration?

Self-Exploration Question

1. Using the following space, list the intervention steps outlined in each case study and describe the benefits and challenges associated with each step. As you reflect

on the challenging parts, list your alternative steps. See Chap. 10 for alternative steps.

Case #1, Ali—Survival Mindset

Case #2, Nimali—Therapeutic Mindset

Case #3, Sarah—Active, Engaged Mindset

Competing Interests The author has no conflicts of interest to declare that are relevant to the content of this chapter.

References

Bandura, A. (1977). *Social learning theory*. Prentice Hall.

Bowlby, J., World Health Organization., & World Health Organization. (1951). *Maternal care and mental health: A report prepared on behalf of the World Health Organization as a contribution to the United Nations programme for the welfare of homeless children*. World Health Organization.

Bowlby, J. (1973). *Attachment and loss*. Basic Books.

Cozolino, L. J. (2010). *The neuroscience of psychotherapy: Healing the social brain* (2nd ed.). W. W. Norton & Company.

Gordon, T., & Borushok, J. (2017). *The ACT approach: A comprehensive guide for acceptance and commitment therapy*. PESI Publishing & Media.

Government of Canada, Statistics Canada. (October 26, 2022). "Ethnic or cultural origin by gender and age: Canada, provinces and territories, census metropolitan areas and census agglomerations with parts". www12.statcan.gc.ca. Retrieved March 26, 2023.

Hadad, M. (2013). *Satisfying life: Positive psychology and personal growth*. Pearson Canada.

Hecker, T., Fetz, S., Ainamani, H., & Elbert, T. (2015). The cycle of violence: Associations between exposure to violence, trauma-related symptoms and aggression—findings from congolese refugees in Uganda. *Journal of Traumatic Stress, 28*(5), 448–455. https://doi.org/10.1002/jts.22046

Johnston, D. (2011). Message from his Excellency the Right Honourable David Johnston Governor General of Canada, on the Occasion of the National Day of Remembrance and Action on Violence Against Women. *Marketwire*.

Kabat-Zinn, J. (2005). *Coming to our senses: Healing ourselves and the world through mindfulness*. Hyperion.

Kübler-Ross, E. (1969). *On Death and Dying (1st ed.)*. Routledge. https://doi.org/10.4324/9780203010495.

Kübler-Ross, E. (1997). *The wheel of life: a memoir of living and dying*. Scribner.

Lawson, J. (2012). Sociological theories of intimate partner violence. *Journal of Human Behavior in the Social Environment, 22*(5), 572–590. https://doi.org/10.1080/10911359.2011.598748

Maciejewski, P. K., Maercker, A., Boelen, P. A., & Prigerson, H. G. (2016). "Prolonged grief disorder" and "persistent complex bereavement disorder", but not "complicated grief", are one and the same diagnostic entity: an analysis of data from the Yale Bereavement Study. *World Psychiatry, 15*(3), 266–275. https://doi.org/10.1002/wps.20348

Mitchell, J. E., & Wupperman, P. (2023). Mindfulness treatments for domestic violence: A review. *Journal of Aggression, Maltreatment & Trauma, 32*(7–8), 1055–1075. https://doi.org/10.1080/10926771.2022.2038752

Ogden, P. (2021). *The pocket guide to sensorimotor psychotherapy in context (Norton Series on Interpersonal Neurobiology)*. W. W. Norton & Company.

Ogden, P., & Fisher, J. (2015). *Sensorimotor psychotherapy: Interventions for trauma and attachment (Csm)*. W. W. Norton & Company.

Porges, S. W. (2007). The polyvagal perspective. *Biological Psychology, 74*(2), 116–143. https://doi.org/10.1016/j.biopsycho.2006.06.009

Porges, S. W. (2017). *The pocket guide to the polyvagal theory: The transformative power of feeling safe*. W.W. Norton.

Schore, A. N. (2012). *The science of the art of psychotherapy: The latest work from a pioneer in the study of the development* (1st ed.). W. W. Norton & Company.

Siegel, D. J. (2010). *The mindful therapist: A Clinician's guide to mindsight and Neural Integration (Norton Series on Interpersonal Neurobiology)*. W. W. Norton & Company.

Van Der Kolk, B. A. (2014). *The body keeps the score: Brain, mind, and body in the healing of trauma*. Penguin Publishing Group.

Chapter 10
Healing Tools

1 Tools for Trauma-Informed, Culturally Based Intervention (TICBI)

The symptoms experienced by many survivors of forced migration who were not born in Western countries highlight that some existing interventions do not align with traditional Western definitions of trauma. As a result, clinical diagnoses or specific intervention tools may not always apply to them. Therefore, when utilizing any of the tools described in this chapter, please refer to Chaps. 4–8 for information on culture, trauma expression, TICBI assessment, and intervention planning. The connection between the body, brain, and mind has been articulated in various ways in this book, illuminating the intricate interplay of thoughts, emotions, and physical sensations that influence the well-being of survivors.

The TICBI recovery process emphasizes identifying the impact of trauma and aims to design interventions informed by cultural and religious contexts. Essential information on planning service needs (Chap. 7) and implementing interventions (Chap. 9), while considering the client's mindset (survival, therapeutic, or active, engaged mindsets), is discussed in prior chapters. In TICBI, each intervention is crafted to restore what was disrupted during each stage of forced migration by reconnecting the survivor's body, brain, and mind through what is known as change talk, often referred to as the "talking cure" (Cozolino, 2010, p. 163).

Engaging in culturally informed conversations with clients fosters collaboration and provides valuable insights into the issues they face, allowing for the design of effective interventions. When seeking services, survivors often grapple with life stressors stemming from trauma responses stored in their memory, particularly in the hippocampus and amygdala (two significant areas of the brain frequently discussed in trauma intervention). Although some of these survival responses can be maladaptive, they remain functional until individuals feel stuck and cannot resolve their problems. For example, those with a survival mindset are primarily guided by

N. Bokore, *Trauma-Informed, Culturally Based Intervention*,
https://doi.org/10.1007/978-3-032-06835-4_10

their adaptive survival behaviors and continue to exhibit maladaptive responses until these severely affect their well-being and quality of life (e.g., Ali in Chap. 9).

In contrast, those with a therapeutic or active, engaged mindset (e.g., Nimali and Sarah in Chap. 9) navigate their day-to-day lives effectively until they encounter an immediate crisis that disrupts their adaptive responses to post-forced migration and emotional balance. In all cases, the effectiveness of TICBI relies on the information gathered during the preliminary assessment (see Chap. 8) regarding adaptive responses, recent crises, symptom expression, values, motivations, and what holds significance for them. Survivors in each mindset continue to lead their day-to-day lives, utilizing solutions stored in their memory banks and matching previously successful strategies to tackle current challenges based on previously structured "somatic markers and engrams," aligning the past with the present (Levine, 2015, p. 130). Since these responses influence both conscious and unconscious actions, the challenge arises for this matching when there is continuous activation undermining survivors' recovery processes and increasing their distress. Some of the struggles for these survivors include the following:

- **Individual challenges arise from** the confusion and struggle stemming from frequent trauma reactions, which can conflict with cultural or religious teachings advocating acceptance of life's difficulties and gratitude for survival. Cultural and spiritual beliefs encompass both benefits and challenges. The benefits derive from a belief system that promotes healing, fostering strength, resilience, and tolerance of past experiences by encouraging believers to accept what cannot be changed. For some, the spiritual challenges they face may exacerbate their trauma, creating feelings of guilt and self-blame rooted in personal inadequacy. Individual challenges are even more pronounced for those with a survival mindset, necessitating a tailored assessment for TICBI that considers the client's experiences, emotions, and belief systems in a safe space. An in-depth assessment/evaluation (see Chap. 9) builds skills to facilitate change, balance emotions, and develop new adaptive behaviors supporting growth and resilience.
- **Interpersonal challenges:** In the wake of the latest personal crisis, survivors across the three mindsets presented in the case studies for this book (Ali, Nimali, and Sarah— Chap. 9) sometimes grapple with interpersonal challenges. For each survivor, a TICBI-trained therapist provides information and tools they can utilize to navigate the immediate crisis (extreme fear, grief/loss, and domestic violence). The work also includes respectfully gathering information about the past and present, offering support by listening to their stories, avoiding assumptions about their culture or religion, and assisting them in reframing their perspectives. Some tools for dealing with interpersonal challenges in TICBI include cognitive processing therapy (CPT) to manage PTSD symptoms, dialectical behavior therapy (DBT) to handle intense emotions arising from current crises, mindfulness practices to learn grounding techniques for emotional balance and enhance sensory awareness, and narrative therapy for reframing maladaptive thoughts, along with other methods that will bolster somatic responses. The aim of TICBI in addressing interpersonal challenges is to assist survivors in processing past

traumatic memories, identifying triggers, and developing coping mechanisms to manage current challenges while effectively regulating their emotions and minimizing interpersonal difficulties. With regulated emotions, survivors learn to interact and communicate with others, seek support, navigate conflict, and cope with grief.

The process of each TICBI technique aims to help survivors of traumatic experiences learn to address the issues affecting them in a safe space and facilitate necessary change. Each intervention skill provides survivors an opportunity for self-understanding and normalizes their reactions (most of them are isolated with their feelings), helping them avoid the disconnection and guilt associated with past maladaptive behaviors and belief systems. Numerous researchers have utilized specialized theories and models to clarify brain function, adaptive responses, and the advantages of talk therapy in trauma intervention (Cozolino, 2010; Levine, 2015; Porges & Dana, 2018; Siegel, 2010, 2011; van der Kolk, 2014). Most existing trauma intervention models focus on Western trauma expression and intervention requirements. However, they can be integrated into TICBI to equip survivors with specialized skills that include self-understanding and working on their recovery by using the cognitive insight survivors gain about their past and present as they reframe their thought processes, reconnect their body, brain, and mind, and promote well-being (Cozolino, 2010).

TICBI-trained social workers begin by informing clients about their reactions and the step-by-step activities facilitating change. Integrating Eastern and Western therapeutic approaches in trauma intervention enables social workers to focus on cognitive processes by evaluating and discussing culturally informed thought patterns and managing emotions, which enhances problem-solving capabilities (top-down). Simultaneously, it helps survivors attend to their physical functions as they learn how their nervous system influences bodily sensations, movements, and posture (bottom-up).

1.1 Bottom-Up and Top-Down Approaches in TICBI

In TICBI, the bottom-up and top-down approaches help address trauma responses and individualized healing practices. The bottom-up approach describes the body's reaction to trauma, emphasizing the nervous system and sensory responses that regulate or dysregulate survivors' emotions. Discussions of the top-down approach, on the other hand, facilitate an exploration of trauma responses and their effects on cognitive processing, highlighting thoughts, belief systems, and emotional regulation strategies. Sharing information about the connection between the body, brain, and mind is vital for culturally informed talk therapy. It often leads to discussions that enhance survivors' motivation and hope for recovery from traumatic memories by engaging the survivor's cognitive processing and conscious thoughts while exploring deeper emotional states (top-down). Similarly, a variety of techniques can be employed to release the physical tension created by traumatic experiences (bottom-up).

Incorporating physical activity and Western methods such as Sensorimotor Psychotherapy in TICBI, a body-centered approach, assists survivors in addressing the impacts of trauma on their body, brain, and mind. While the inclusion of the Eastern approach, such as yoga, meditation, and cultural dance, contributes to the healing process. For instance, if a client self-isolates due to fear, focusing on understanding the source of that fear, reframing, rebuilding social connections, and participating in social or cultural activities for support is a strong starting point and pathway toward long-term recovery. Healing begins with discussions involving top-down activation, such as the neocortex and hippocampus, while facilitating the analysis of the presenting issue, creating an opportunity for self-reflection and reframing thoughts. Recovery comes with increased survivors' insight into their encoded memories of the past, helping them develop new perspectives that lead to recovery.

Likewise, encouraging physical activities in the healing process enables survivors to examine the impact of problematic emotions on their physical sensations and the connection to their emotional responses (bottom-up). Therefore, the benefits of both top-down and bottom-up activation in TICBI are an integrative approach to assist survivors in recovering from their past experiences while fostering resilience and growth (Ogden & Fisher, 2015; Ogden et al., 2006). For instance, it would be beneficial for survivors like Ali to comprehend the impact of his reaction (fear and self-isolation) as bottom-up sensory information from his brainstem at the lower layer of his cortex and how these neuronal activities transmit sensory information to his neocortex for cognitive processing. This meaningful information about his reaction and impact helps him understand his post-displacement reactions and actions. In TICBI, integrating top-down and bottom-up activities enhances survivors' responses as they move forward and heightens their interoceptive awareness to foster lasting change (Siegel, 2010).

The human brain also functions beyond merely connecting top-down and bottom-up communication; it interlinks horizontally between the left brain (logical thinking) and the right brain (emotional processing) (Siegel, 2007, 2010, 2011). These connections play a crucial role in the healing process in TICBI, allowing survivors to engage in critical thinking that enhances not only vertical connections (top to bottom or bottom to top) but also horizontal connections (left and right brain). These multiple links support survivors in their healing process as they learn to deactivate their primal or survival instincts and responses, thereby strengthening their ventral vagal reaction (increasing the activities of the social brain) and re-engaging brain areas previously conditioned to survival reactions or responses.

Sensorimotor therapy (discussed below) is a culturally congruent therapeutic technique involving vertical and horizontal brain activation with the increase in their awareness of their physical sensations and emotions as they participate in activities designed to release trauma-related physical tensions. Vertical integration mainly includes vestibular stimulation[1], enhances the "downward" connection (vestibular function), and boosts the "upward" cognitive activity and connectivity, which improves sensory input motor control and positively influences higher-level brain functions. Similarly, horizontal integration during sensorimotor therapy strengthens the connection between different brain regions and hemispheres, integrating sensory information with motor actions. These activations are enhanced in the talking cure process when the intervention is combined with a culturally accepted and accessible healing tool, such as dancing to a favorite song for therapeutic purposes. For instance, I have utilized the Ethiopian *Eskista* (a traditional dance) as an enjoyable cardio activity to alleviate stress and symptoms of situational depression. Yoga or stretching exercises are another culturally acceptable form popular among urbanized and educated survivors, which I have employed in the past for individuals dealing with sensory-motor amnesia (SMA) that affects survivors.

Influenced by Hans Selye's (1976) general adaptation syndrome, Thomas Hanna (1986, 1988, 1990), a pioneering visionary in somatic studies and research, introduced the concept of sensory-motor amnesia (SMA), which often occurs in survivors of prolonged trauma. This condition refers to muscle contractions stemming from a chronic stress reaction or extended traumatic experiences that disrupt the autonomic nervous system's response, frequently causing physical pain due to a disconnect between the body and brain. In these instances, a bottom-up approach, such as yoga, alleviates the chronic muscle tension and pain caused by the repeated experiences of hyper- or hypoarousal, which impede sensory-motor function and overall well-being. Both cardio and stretching activities promote the flow of horizontal energy, blood, and the connection between the body and brain, aiding in trauma recovery and enhancing communication among the prefrontal cortex (PFC), somatosensory cortex, and hippocampus, thus playing a role in trauma processing and recovery.

Trauma intervention that includes repeated physical activity, such as the Ethiopian Eskista dance for cardio exercise, yoga for stretching and centeredness, or other preferred activities for emotional and physical pain, allows survivors to reconnect their autonomic imprints that, over time, rearrange the disconnection and create new adaptations and autonomic responses (Hebb's Rule, 1949). The creation of these new mental imprints generates new memories through reframing. It helps develop new habits that transition survivors from a survival mindset to a therapeutic or actively engaged one (see Chaps. 8 and 9). With continued practice of top-down and bottom-up approaches to healing, these new neural links will continue to strengthen and help survivors cultivate new autonomic responses and promote post-traumatic growth-based narratives.

Every preferred activity undertaken during talk therapy increases the chances of your client's neural integration, reconnection, and change. For example, during a narrative therapy session, following the bottom-up approach and using specific techniques such as deep breathing and meditation helps decrease implicit responses

---

[1] Input from physical activity such as movements or gravity the body recives.

often triggered by the danger alerts from the lower subcortical area (including the fear circuit/amygdala). Integrating deep breathing or meditation with culturally and religiously acceptable practices helps survivors stabilize their minds, refocus their attention, and repair their traumatized and scattered brains, improving their thought processes and concentration. Physical activity decreases stress and increases mental tranquility, widening their tolerance and promoting well-being (Ogden & Fisher, 2015; Siegel, 2007; van der Kolk, 2014).

Similarly, enhancing sensorimotor activities to alleviate SMA through stretching (yoga) and other methods like Pilates and tai chi offers a culturally relevant intervention to help survivors escape the emotional turmoil and physical pain they endure, irrespective of their mindset. Activities that establish a bottom-up and top-down connection between the body and brain assist survivors in boosting their sensorimotor awareness and overall well-being. I have employed these activities to lessen my own trauma-related tension or mitigate the effects of a trigger, which prompted me to incorporate them into my intervention plans. Client feedback has been outstanding, and most importantly, witnessing their benefits from the activities and their growth in sensorimotor awareness and recovery reinforced my faith in these methods, making them integral to my TICBI approach in this book. I found it advantageous to outline accessible recovery activities that fuse body/brain connections, reframe my clients' narratives, and mend their wounds by reconnecting the hippocampus (responsible for learning, memory, and storing new adaptations) and the prefrontal cortex (which alters thought processes in the thinking center).

1.2 Social Work Theories, Body/Brain and Mind Connections

Integrating brain/body conversations using neuroscience knowledge (body/brain and mind talk) for TICBI enhances talk therapy by providing evidence-based information acceptable to most cultures. It fosters a deeper understanding of the biological underpinnings of trauma and the impacts of body/brain responses on survivors. Additionally, it facilitates the collaborative development of effective TICBI, addressing both the emotional and physiological effects of prolonged trauma and the healing process. For example, the narrative approach introduced in Chap. 7 enables survivors to reflect on their experiences, assess the impact of their past encounters, and discover new pathways to healing through reframing, engaging survivors in cultural communication about the power of human imagination to heal the body, brain, and mind, thereby increasing their brain's vertical and horizontal connections. This neuroscience-based information sharing can begin with a simple question that starts with "what if?" to assist survivors in evaluating their thoughts, actions, and imagination, altering their perspective, and fostering upper subcortical activation (based on top-down strategies). This process enhances the activation and connection between the prefrontal cortex, the insular cortex, and the hippocampus, allowing survivors to achieve interoceptive perception and comprehend what is occurring within their body/brain to facilitate the necessary changes. This talk therapy method, combined with brain activation during narrative re-storying (reframing) or retelling their experiences, helps evaluate their struggles from various perspectives while making necessary adjustments. The change process is challenging for survivors, stressing the need for having an established rapport (using the person-centered approach) and trust with survivors to engage in the suggested healing process, including the additional physical activities aimed at alleviating symptoms of SMA.

1.3 The Person-Centered Approach

The person-centered approach (sometimes referred to as the client-centered approach; Rogers, 1951) helps provide personalized care to survivors based on their needs, wants, and interests. In TICBI, this approach emphasizes cultivating a safe, supportive, and empathetic environment that allows for collaborative exploration of the survivor's experiences and emotions, thereby promoting the self-healing process and personal growth. The approach encourages starting the healing process by alleviating survivors' tension and the stress associated with learning new activities while strengthening the relationship between the social worker or therapist and the client. It facilitates trust, allowing survivors to share their stories in a safe space or middle-layer level (see Chap. 9, Fig. 9.1). It also establishes a secure environment for engaging survivors in mindful conversations that integrate their body, brain, and mind. Ultimately, TICBI therapeutic skills informed by this approach are designed to enhance survivor self-awareness and the social worker's practice of attuned listening for co-regulation (Applegate & Shapiro, 2005; Dana, 2020; Farmer, 2008). Employing theories like this encourages survivors to engage in the change process within a safe space, thereby increasing their window of tolerance and promoting well-being (Ogden, 2021).

In TICBI, the person-centered approach allows social workers to concentrate on the role of culture in their clients' descriptions of internal and external stimuli and their subsequent responses. Trauma intervention involves treating survivors with dignity, respect, and compassion while empowering them to share their survival narratives and value-based perspectives about the issues affecting their well-being. Judith Herman (1997) described this approach as part of "the first principle of recovery" (p. 190) in trauma intervention. The person-centered approach aids in the recovery process as survivors regain their strength, reduce stress, and enhance their self-confidence. This approach strengthens the working relationship, fostering trust, a crucial aspect of trauma intervention.

Talk therapy based on a person-centered approach also aids in deactivating specific brain areas, particularly in the left hemisphere, which is responsible for language and logical thinking. Simultaneously, it activates the right hemisphere, a source for processing non-verbal emotions, and the medial prefrontal cortex (mPFC), which enhances affect regulation, introspective awareness, and resilience. Talk therapy engages the left hemisphere, including the medial prefrontal cortex (mPFC) and the anterior cingulate cortex (ACC), found in both the right and left hemispheres, which are crucial for emotional regulation and cognitive control. Survivors, particularly those with PTSD symptoms, often exhibit reduced activity in both the mPFC and ACC, hindering their ability to recall traumatic memories or process emotions due to memory avoidance and hypoactivity.

Another benefit of using a person-centered approach in TICBI is that the conversation helps survivors identify their challenges and strengths in a safe space, guiding them toward recovery by pinpointing resources and leveraging their capabilities to facilitate change. The combination of activating the medial prefrontal cortex (mPFC) and the anterior cingulate cortex (ACC), along with trust-building and collaborative relationships during the talking cure, further enhances survivors' neural reconnection in these and other areas involved in affect regulation, including the prefrontal cortex (anterior and medial regions). The medial prefrontal cortex (mPFC), comprising the ventromedial (vmPFC) and dorsomedial (dmPFC) areas, plays a crucial role in self-awareness, emotional processing, and social cognition, all of which are essential for introspective awareness.[2] Internal bodily sensations and emotions are influenced by cultural learning (see Chap. 6), which shapes how people behave, perceive, and process bodily sensations and emotions, affecting one's interoceptive awareness. Therefore, culturally based communication and the person-centered approach aid in recognizing survivors' strengths and core values, driving the change.

1.4 Focusing on Personal Strength During TICBI

TICBI focuses on personal strength to continue healing, allowing survivors to engage in resilience-based, action-oriented conversations. Emphasizing personal strength helps identify resources available to survivors for reciprocal change talks, where their strengths are recognized and their self-determination is respected. The change talk activates the brain region involved in interoceptive sensations (Siegle, 2007), which is crucial for understanding physical and emotional states (from the anterior and posterior insula). Concentrating on personal strength enables collaborative and evaluative discussions to reframe maladaptive narratives and subjective meaning-making processes.

The approach helps survivors realize their strength and ability to manage their feelings, sensations, and emotions while making the necessary changes. It enables survivors to observe and evaluate their thoughts, feelings, and mental processes (interoception). This activity enhances the insula's transmission of sensory information from the body to the brain, helping survivors "look inward" to assess past learning about what needs to change to promote well-being. Looking inward for

[2] The ability to be aware or understand one's own internal states, including thoughts, feelings, and motivations.

meaning-making is "inherent in every culture's contemplative practice" (Siegle, 2007, p. 103), making it an acceptable intervention process in providing TICBI.

Various theories are used in TICBI and are intended to help those disempowered by war and forced migration to regain their power, build confidence, and increase body/brain connection and cognitive function, promoting internal (sensory) and external (physiological) changes. However, each intervention step must be undertaken carefully for clients, considering their survival mindset. Conversations about hope and self-empowerment are initially threatening because they challenge the survivor's active survival responses (fight, flight, or freeze), and the survivor might abandon the change talk.

Work with survivors, considering their present strengths and resilience, gradually introducing each activity, allowing them to participate comfortably, motivating them to work toward their healing journey, and moving away from the pessimism, powerlessness, and apathy that overwhelm survivors, particularly those in survival mindsets. Therefore, it is crucial in TICBI to let survivors determine their pace and preferred intervention methods. Openness and flexibility during change talks enable survivors to express their concerns and exchange knowledge, especially regarding taboo subjects or other potential stressors, reactivating past traumas and triggering sympathetic autonomic responses. This approach is an integral part of TICBI. It aims to help survivors recognize their internal strengths while providing empowering opportunities to reframe their narratives and transform their physiological and psychological responses to trauma.

1.5 Strength Through Reframing Narratives

The narrative approach to trauma recovery has been recognized and utilized since Michael White and David Epston introduced it in the 1990s. Therefore, reframing narratives of maladaptive survival actions is essential for facilitating change. It enables using personal strength to explore survivors' values, belief systems, and fears, reshaping their stories with hope narratives emphasizing their strengths and resilience. Reframing narratives in TICBI benefits survivors across various mindsets. For instance, (1) it assists individuals dealing with prolonged emotional dysregulation, disconnection, protection-based sympathetic activation, and dorsal vagal shutdown (survival mindset), and (2) it supports survivors who otherwise function well but are navigating unexpected crises that trigger their past experiences and activate their sympathetic mobilization system or "fight-or-flight" response (see Chap. 9 for therapeutic mindset or active, engaged mindset).

Reflecting on the case studies in Chap. 9, employing TICBI allows survivors like Ali to reshape their narratives with culturally and religiously informed values and motivations, moving beyond post-war, problem-saturated tales of defeat that lead to sleepless nights, feelings of anger and revenge, and a sense of helplessness. For survivors like Ali, who have endured isolation while coping with these symptoms and facing barriers for an extended period, the initial application of TICBI-based activities can be daunting. For him, sharing his personal feelings with strangers or the social worker will likely remain limited to a surface-level conversation (see Chap. 9, Fig. 9.1). The limitation for the initial conversations may involve understanding his self-defeating language, justifying his self-isolation: "I cannot sleep. I have been contemplating what people say about me. I cannot attend the mosque because my financial situation is embarrassing. Seeing too many people reminds me of the past." Reframing his thoughts and feelings requires establishing trust, which is crucial for eliciting a response to the guided questions regarding his concerns, identifying his strengths, providing him with additional information to normalize his reactions, and reframing his thoughts. For survivors like Ali, after establishing trust, reframing his narratives may include the use of the following recovery-based guided and imaginative questions or discussions.

- Imagine that after visiting the mosque and participating in the prayer, you encounter someone who turns out to be the best friend you thought you had lost in the war.
- What if the person next to you during *salah/salat* becomes your friend?
- What if your new friend is someone who can provide job opportunities and improve your financial situation?

These questions and guided conversations aim to reduce stress, build relationships, and encourage Ali to view the issue of self-isolation and self-blame from a different perspective, reframing his thoughts and helping him avoid distressing thought patterns, including discussions of taboo subjects. Fear often traps survivors like Ali in negative thinking patterns of "I cannot" (bottom-up activation), and shifting to a "what if?" thought process (top-down activation) facilitates reframing and neural reconnection.

For some survivors, these conversations may trigger their sensory-motor amnesia (SMA), which entails a loss of awareness of physical reactions, including dysfunctional patterns of body movement, posture, and muscle memory, leading to pain. Therefore, incorporating breathing exercises is crucial for alleviating physical and emotional responses. Reframing maladaptive narratives to decrease physical and emotional reactions is necessary. To reduce the stress of reactive responses, it is advisable to introduce questions and activities gradually, step by step, to prevent undue distress and re-traumatization. Experienced social workers employ various skills to ease tension and recognize distorted thought patterns, helping survivors identify alternative thought processes and manage their physical reactions. Beneficial activities for gaining strength and reframing thought patterns include creating a chart to pinpoint the first cognitively distorted views affecting the thought process and encouraging consideration of alternative perspectives. For example, Ali's chart may resemble the following:

Cognitive Distortion for Individuals with a Survival Mindset:

- I take responsibility for everything that happened to my family during the escape. As a man, I should have stayed close to my family to protect them.

Questions For Alternative Thinking and Reframing Approach:

- Was staying with them an option for me back then?
- What more could I have done to prevent what happened?

Reframing questions like this helps survivors assess their thought processes and actions, reflect on their belief systems, and identify external sources that challenge those thoughts (e.g., during the war, men often face the choice of killing or being killed). Recognizing these external forces that limit Ali's options and value-based distortions affecting his thinking enables him to explore alternatives to re-author his narrative. This change reduces sympathetic activation and enhances access to his cognitive processes, empowering Ali with new perspectives, an awareness of possibilities, and a recognition of his strength and resilience.

Another benefit of focusing on strength-building through reframing narratives is that it enhances the survivor's ability to engage in additional therapeutic activities that help reduce the physical tension that may arise. Some of these activities include pendulation and titration techniques, which are part of somatic therapy used to assist individuals in processing traumatic memories along with their related emotional and physical responses in a safe space (Levine, 2015). Pendulation and titration techniques provide an approach to release trauma-related sensations and emotions for recovery.

1.6 *Emotional Regulation Through Pendulation and Titration Techniques*

1.6.1 Pendulation

Peter Levine (2015, p.55) first proposed the use of pendulation and titration in trauma intervention as fundamental concepts in somatic experience (SE) to address the effects of implicit traumatic physical sensations and emotional memories. For instance, during the previously mentioned reframing activities, pendulation techniques can involve paying attention to conversations that evoke physical sensations of discomfort and memories related to cultivating safety and calm. The intervention includes the social worker asking questions that gradually shift the survivor between survival stories, inciting physical sensations and emotional arousal while reframing these to foster calm and relaxation.

During pendulation activities, incorporating therapeutic support and security helps survivors learn to regulate their feelings and emotions, minimizing overwhelming physical sensations in their nervous systems. Exploring past traumatic experiences and reframing the narrative can be stressful for some survivors, as it triggers memories, leading to a heightened state of arousal or hypervigilance. Some techniques discussed below, such as breathing or grounding exercises, can be included to contribute to the calming process.

In somatic therapy, the pendulation technique involves a slow-paced dialogue emphasizing engaging in conversations and activities connecting the body, brain, and mind as survivors learn to pay attention to their maladaptive thought processes and implicit physical responses. These activities are designed to help survivors manage their emotions as they transition from trauma, anxiety, and fear, which cause physical contraction and tension (e.g., faster breathing, increased heart rate, and muscle tension) to a state of recognition, release, and relaxation.

The techniques enable survivors to recall memories that may trigger hyperarousal or hypoarousal (amygdala activation) responses in a safe space, allowing them to renegotiate or reframe their reactions and perspectives on the event, thus shifting their emotional responses toward the center of the window of tolerance. As mentioned earlier, the effectiveness of this activity relies on trust, collaboration, and emotional co-regulation, with the social worker facilitating the conversation, pacing the activities, and incorporating breathing techniques as necessary to help the survivor balance their emotions and physical responses (through top-down and bottom-up connections) while remaining anchored in the present moment.

The activity should begin by discussing the least distressing experience related to forced migration, encouraging re-evaluation of the thoughts and actions connected to those experiences.

TICBI therapists use pendulation to enhance the effectiveness of reframing narratives, helping survivors increase their hope for healing and providing them with the strength and motivation to confront future challenges with renewed perspectives. They facilitate change by actively assessing narrative interpretation, recognizing its impact, and building a positive and empowering account of the event that alters their responses. In TICBI, working with survivors and asking them about their past requires a delicate approach. Taking each activity slowly and allowing the conversation to build gradually promotes healing. Talk therapy using TICBI necessitates patience for the survivor's nervous system to adapt over time, deactivating the amygdala and sympathetic response to continue the healing process in a more comfortable space.

When providing TICBI, it is necessary to pay attention to navigating emotional regulation and dysregulation, as well as transitioning between amygdala activation and deactivation, along with the associated challenges. This is particularly important when working with individuals with a survival mindset (e.g., Ali), as their emotions often remain unstable. This instability may increase the likelihood of withdrawal from TICBI due to fear. Therefore, a crucial first step is implementing slow-paced, introductory activities to help survivors re-regulate their dysregulated emotions. This includes beginning with psychoeducation to assist survivors in understanding their feelings, normalizing their reactions related to trauma, and gradually moving away from the stuck points that restrict their emotions, pushing them away from their window of tolerance toward hyperarousal or hypoarousal states.

Like pendulation, using "titration" techniques is a gradual, step-by-step process designed for survivors to recall tucked-away stressful events or experiences and release the physical tension and emotions tied to them. The titration techniques allow collaborative analysis of these experiences in a safe environment without overwhelming or retraumatizing survivors. Titration activities diminish or modify trauma-related somatic experiences and implicit or procedural memories from the past. However, these activities can be challenging and may sometimes overwhelm survivors due to the implicit reactions of an overactivated amygdala.

1.6.2 Titration

While pendulation involves activities that move back and forth between sensations of safety and discomfort, titration techniques, on the other hand, break overwhelming experiences and related emotions into smaller, manageable parts. The titration technique is often mentioned in non-social work disciplines; for instance, in chemistry, it is defined as the method used to determine the intensity of a substance when diluted with a known concentration (the titrant), gradually diluting the intensity to indicate a change. In TICBI, "titration" is described as another somatic therapy technique used to help survivors address intense emotions caused by distorted thought processes and related somatic responses.

Therefore, in trauma therapy, titration techniques mainly aim to challenge or "de-potentiate" maladaptive narratives and implicit reactions associated with procedural memories that cause physical or somatic pain (Levine, 2015, p. 132). The titration technique is a gradual process of recalling complex memories and related emotions for reframing and increasing interoceptive awareness, which helps to reduce the intensity with calm and possibilities. It enhances self-awareness of internal states, feelings, and sensations to alter intensity and support healing (Levine, 2015).

Practical activities include gradually remembering fear-based survival stories, step by step activating the amygdala, and learning to relieve tension through breathing or relaxation techniques. These techniques enable survivors to process their traumatic past and related emotions in a controlled and supportive environment, helping them learn to manage their dysregulated emotions and prevent the overactivation of their amygdala (fear response). Titration activities can assist and motivate clients to learn how to process memories in a safe space while challenging their reactions to their experiences, using small, manageable steps at a slow pace to build tolerance, renegotiating trauma, and avoiding the impact of overwhelming emotions.

1.7 Integrating Pendulation and Titration Techniques In TICBI

Integrating pendulation and titration is beneficial for TICBI, as it aids in altering the survivor's body and brain responses while managing their trauma reactions by acknowledging and releasing related sensations and emotions. For instance, combining these techniques establishes a "top-down" connection involving the prefrontal cortex, engaging the mind and brain to promote emotional regulation. Likewise, a "bottom-up" activation normalizes trauma responses and enhances the regulation of physical sensations and emotional reactions for survivors, particularly those experiencing stuck points of survival (survival mindset). Incorporating these activities helps survivors learn to manage their emotions and distress responses that impact their bodies, brains, and minds.

Since these activities involve activating and deactivating the amygdala, which can be stressful for survivors, integrating mindfulness practices with these techniques helps them adapt more effectively to what they learn, providing opportunities for self-reflection and evaluation of their thoughts and actions. Including these techniques in TICBI also highlights the importance of raising awareness of the value-based purpose of life that many cultures embrace while fostering nonjudgmental thoughts to help understand their post-trauma emotions and actions as they arise. Gaining insight into self-empathy while focusing on strength and resilience empowers survivors to let go of the self-judgment and fear that often keep them stuck. Through practices based on pendulation and titration, the chances of recovery increase as survivors engage in new adaptations and changes, allowing them to gain fresh perspectives, develop new narratives, improve their self-confidence, power, and control over their lives, and experience the benefits of a shift in their hierarchical autonomic response (from a survival mindset to an active, engaged mindset).

Helpful tips for these activities include paying attention to reactions during the gentle exploration of traumatic experiences (titration) or trauma responses in a safe environment (pendulation), and checking whether the activity increases or decreases heart rate. If the heart rate has increased, introduce breathing techniques such as diaphragmatic breathing to help relax both the body and mind. This breathing practice, which focuses on the movement of the stomach rather than the chest (expanding and contracting the stomach) during inhalation and exhalation, directs the survivor's attention to their breathing, enhances physical awareness, and distracts them from stressing over difficult emotions. Additionally, this technique also emphasizes each breath, increasing the likelihood of relaxation and fostering a calm mind.

When providing TICBI, the integrated activities described in this chapter aim to help clients develop their skills in healing by learning to regulate emotions, especially when past adaptive responses make managing current crises difficult. However, each activity must be suitable and flexible to meet the client's needs, ensuring it is culturally and spiritually appropriate. Another important point to consider when using pendulation and titration activities, or combining both, is to recognize the client's varied reactions and adjust accordingly to support ongoing recovery from the intervention.

2 Other Integrative Approaches for TICBI

Other integrative approaches in TICBI include culturally and religiously appropriate methods, such as somatic mindfulness practices that relieve physical and emotional tension or pain. Sensorimotor therapy addresses this method, looking at the manifestations of trauma in the body and encouraging movement-based, culturally informed healing techniques to improve body–brain awareness and recovery. The integrated approaches employ slow and deliberate therapeutic methods involving mindful movements, such as yoga, tai chi, or Pilates, to enhance understanding and strengthen the mind–body connection, alongside healing practices grounded in breathing exercises. In TICBI, integrative approaches significantly contribute to the client's healing journey by employing the previously mentioned techniques, as well as others. For instance, (1) cognitive-behavioral therapy (CBT) can be paired with narrative therapy to reframe unhelpful thinking patterns such as self-blame and fear-driven responses, while (2) psychodynamic therapy can help to explore unconscious survival responses that affect behaviors and sensorimotor reactions.

Integrating these approaches into the client's healing practices or the trauma recovery process is not new in social work. However, what currently stands out as innovative in the field is the application of neuroscientific knowledge as a valuable tool for educating clients and enhancing survivors' understanding of their actions and reactions that contribute to the crises in their lives. In TICBI, each approach is tailored to client needs, providing foundational information about the issues, recovery-related practices, and healing benefits. For instance, when working with clients entrenched in a survival mindset, such as Ali (see Chap. 8), who struggles with a lack of balance—characterized by a constantly activated amygdala and hyper- or hypoarousal states—that affects his daily life and causes intense immobilization and social withdrawal, using CBT or psychodynamic therapy along with straightforward narrative therapy to explore his unconscious survival responses and reframe his maladaptive thought processes will be beneficial. This approach will empower him to shift his self-blame and isolating thoughts toward an action-oriented and hope-driven healing path, connecting his brain, body, and mind to work toward positive change.

In social work, we refer to integrative approaches as "eclectic practices," which emphasize using various techniques in talk therapy to enhance therapeutic impact and facilitate recovery. Eclectic practices, particularly in trauma-informed intervention, enable social workers to introduce therapeutic activities that not only reframe survival narratives but also rewire survivors' dysregulated autonomic nervous systems, promoting well-being. Utilizing simple, accessible, and acceptable techniques, such as mindful breathing, can help alleviate SMA symptoms through sensorimotor therapy involving body movements, thereby supporting recovery.

These activities are vital in TICBI due to the connection between the body, brain, and mind. For instance, the mind simultaneously holds and manages numerous thoughts (including those constrained by taboo), emotions, and experiences that are typically regulated. However, traumatic experiences can disrupt this process, creating dysregulation that overwhelms the body, brain, and mind's ability to process life events, heightening trauma responses such as PTSD symptoms (e.g., anxiety, anger, sadness, sleeplessness, or self-isolation).

Furthermore, some survivors may experience narrative amnesia, which impairs their ability to create coherent stories about their past or present and influences the therapeutic process. For these survivors, the disconnect between the body, brain, and mind caused by traumatic events is apparent, especially when they are dealing with recent crises (see Chap. 8 for Nimali's and Sarah's case studies). In these cases, they struggle during talk therapy, as their bodies and minds sometimes fail to express their feelings or emotions clearly.

Peter Levine (2015) explains these humanistic somatic experiences (SE) by illustrating how various animals recover from traumatic encounters. For instance, prey animals quickly recover from potential attacks by undergoing a physical process that releases the nervous energy built up during the threat. In contrast, humans do not experience this immediate physical release, causing their traumatic experiences to persist in their minds and leading to inability to express their feelings, anxiety, self-blame, and other detrimental reactions such as self-isolation and anger. Levine suggests that somatic-based therapeutic techniques help survivors by offering them a route away from the emotions that confine them and moving them toward healing.

I recall grappling with past emotions, such as anger toward those who hurt me, sadness over the significant losses in my life, and fear about how I would manage to raise my four children in a new country while simultaneously rebuilding my career. I remember attending yoga classes at the gym, where the instructor taught mindful breathing exercises to help relax and release tension at the end of each session. Those simple stretching and breathing exercises, guided by the instructor, helped relieve some of the physical and emotional stress I had carried for years. After participating in the first few classes, I noticed a significant change in my physical tension and a difference in how I felt. Since then, this practice has become an integral part of my life; I often rely on it when emotions related to my past wounds resurface. It continues to help me recognize my feelings, acknowledge the pain, and center myself in the present moment with increased self-compassion and decreased self-blame. This physical activity still reminds me to let go and choose to inhabit the present rather than dwell on the sadness and anger of the past. Mindfulness practice, as introduced in Chap. 9, serves as a valuable recovery process for survivors of war and forced displacement; mindful breathing techniques can soothe emotions anywhere and anytime without incurring any costs or guilt from religious restrictions.

2.1 Mindfulness Practices for Recovery

The connection between the body, brain, and mind is a universal gift that highlights the need for holistic integration of all three for survival. It outlines the complex interaction among thoughts, feelings, bodily sensations, emotions, and overall well-being. Introducing mindfulness into TICBI practice arises from these reflections, my personal experiences, and my work with survivors. Therefore, incorporating mindful breathing techniques when addressing specific aspects of trauma—using methods accepted across all cultures and religions—promotes this connection and the importance of paying attention to the physical, mental, and emotional responses linked to traumatic experiences and their effects on the survivor's well-being. This allows survivors to see sensations of fear as temporary emotions, rather than getting trapped in them, helping to manage maladaptive reactions and support change. Mindful breathing paves the way for transformation through self-awareness, fostering calmness, emotional flexibility, acceptance, and ultimately helping to dispel fears.

In TICBI, when introducing mindfulness practices and other breathing techniques to culturally diverse groups, it is essential to provide information on the healing aspects of each practice. This helps determine the acceptability of the activity to the client. For instance, yoga, meditation, and mindfulness are often associated with Eastern traditions, and some clients might resist their use. By presenting them as relatable methods employed by various cultures and religions to foster calm, emotional balance, and centeredness, we can increase the likelihood that clients will try and experience the benefits. The key to introducing somatic mindfulness begins with explaining how these practices enhance insight into thought patterns and actions, revealing unconscious thoughts and emotions and their effects on well-being. Since these activities can be unfamiliar to some survivors, it is advisable to start slowly with simple breathing techniques to gauge their response and observe the benefits of promoting emotional calm. The gradual introduction and practice can motivate clients to engage more fully in the talking cure process and other methods to alleviate emotional and physical tension from their past. In this journey, for example, incorporating cognitive-behavioral therapy (CBT) or psychodynamic therapy can assist in moving those who are trapped in culturally informed preconceptions of "should" and "ought to," which can significantly cause tension/pain and impact their well-being.

2.2 *Meditation for Recovery*

Meditation is another tool that can be used by social workers practicing TICBI. The practice offers an opportunity to cultivate centeredness and foster self-compassion, which often eludes trauma survivors. It facilitates the emergence of ventral vagal energy and the importance of connections. A notable meditation that activates the ventral vagal system is the ancient loving-kindness meditation, also known as metta meditation, which focuses on self-generated feelings of love, compassion, and goodwill toward oneself and others—feelings frequently absent for trauma survivors. This meditation practice, linked to Buddhist traditions, also appears in other religions with similar objectives and diverse methodologies (Siegle, 2007).

Loving-kindness meditation nurtures compassion and empathy by inviting individuals to direct positive thoughts and wishes toward themselves and others. The practice begins with self-affirmations for survivors: "May I be safe and protected from harm. May I live without pain and find peace of mind; may I let go of the hurt and learn to forgive, ultimately learning to be at ease and improve my mental well-being" (Siegel, 2007, pp. 62–63). These affirmations create a ventral vagal connection, enhancing the mind's awareness of the body and the heart beneath the chest as survivors learn to breathe more easily (Levine, 2015; Siegel, 2007).

Incorporating meditation and other mindful practices into TICBI helps survivors reduce stress from triggers, enhances emotional resilience, activates their ventral vagal nerve, and improves overall well-being. However, it requires awareness of the client's beliefs and knowledge. Various cultures and religions recommend diverse methods for relaxing and centering themselves after traumatic or stressful life events. Meditation encourages looking inward with focused attention, redirecting thoughts to the present moment. Although it can be initially challenging for survivors of prolonged trauma, particularly those stuck in the survival mindset (see Chap. 9 case study Ali), these techniques can help them work through their anxiety and other trauma responses. Meditation allows survivors to detach from maladaptive thoughts and explore new ways of thinking, thereby allowing them to reconnect with their inner selves.

Research indicates that a short, daily practice of meditation (13–15 minutes) improves attention, memory, and mood, helping to regulate emotions (Ogden, 2021; Siegel, 2007, 2021). This practice aids survivors in recovery as they learn to recognize moments of inner strength and resilience, enhancing their self-confidence and hope. It further helps them observe their thoughts and emotions in the present moment without judgment or reactivity to external stimuli, which those dealing with PTSD often experience (Ogden, 2021; Shapiro & Shapiro, 2019). Additionally, it supports clients in detaching from their feelings, sensations, and thought processes, which is essential for achieving mental clarity and emotional balance.

The practice of meditation has helped me work with survivors. It has benefited me personally, as I have relearned to center myself by stabilizing my emotions when life situations trigger me. It made me aware of the chronic and debilitating emotional and physical pain I have dealt with for years. It facilitated my transition from teaching fitness to practicing clinical social work. Years later, I still practice mindfulness, meditation, yoga, tai chi, and Pilates as part of my self-care and ongoing recovery. Reading more of Kabat-Zinn's work, along with Daniel Siegel's *Becoming Aware* (2021) and Ed and Deb Shapiro's (2019) *The Unexpected Power of Mindfulness and Meditation,* has helped me understand the connection between body, brain, and mind. As a result, I continue to enjoy a life of post-traumatic growth (PTG—see Chap. 4) and live in the present moment. As a survivor of multiple traumatic events, I incorporate what I learned from cultural and religious teachings, such as using prayers for meditation, centeredness, hopefulness, self-compassion, and forgiveness, into my daily mindfulness practice. Numerous religious and cultural groups teach mindful meditation, using their practice methods to emphasize the present moment. Therefore, introducing the practice to clients provides a long-term healing technique and extended recovery, benefiting future generations (epigenetics) by reducing the impact of intergenerational trauma.

Social workers practicing in a multicultural setting must continually deepen their understanding of the connections between the body, brain, and mind, along with the healing roles of culture and religion when providing trauma-informed, culturally based intervention (TICBI). Promoting well-being begins with social workers expanding their cultural knowledge (see Chap. 8) and enhancing their eclectic practices. For instance, as a survivor of war-related trauma, my ongoing recovery involves using deep breathing techniques after morning prayers and meditation to center myself and prepare for the day's possibilities and challenges. After explaining how trauma disrupts natural calm and centeredness by activating survival-oriented protection systems—often characterized by holding one's breath, experiencing improper breathing, or chronic muscle tension—I provided survivors with a copy of the guided script as a take-home activity, especially for those living in triggering situations. Drawing on my experiences practicing these activities and working with clients for over two decades, I integrated them into my social work practice. I recorded 5- to 10-minute meditation exercises, which I share with clients to assist them in practicing the techniques for recovery.

2.3 *Grounding Techniques for Sensorimotor Connection and Recovery*

Recovery through grounding techniques for trauma begins with understanding the biological responses to trauma, particularly the role of the autonomic nervous system in regulating the flow of energy and information among the brain, body, and mind—all of which traumatic experiences can significantly disrupt. To shift the adaptive autonomic responses that survivors develop for protection, they can engage in activities or techniques that promote re-grounding, self-soothing, and resilience, enhancing their capacity to feel safe and re-establish connections with others. Two beneficial activities for recovery include deep breathing techniques, which help survivors regain a sense of being in the present moment and experience the sensorimotor connection. These methods can recalibrate the nervous system and transition individuals from a heightened state of arousal to a more regulated and grounded state.

For instance, survivors may initially find sensorimotor breathing and relaxation techniques challenging due to implicit danger signals in the background. However, practicing beneficial techniques that address disconnection and the painful effects of trauma, especially when dealing with dysregulated sensorimotor responses or other maladaptive behavioral reactions, is essential for providing TICBI. Starting slowly with simple breathing activities helps overcome the complex reactions to traumatic experiences. It encourages survivors to relearn how to relax and breathe normally, which activates their parasympathetic response. Incorporate modest deep breathing and grounding techniques to assist survivors in focusing on the present moment, calming their nervous system. Breathing activities target sensorimotor responses to facilitate self-regulation and manage emotions (Seigle, 2007 & 2021; Ogden & Fisher, 2015; Ogden et al., 2006). For example, I use some activities to engage clients in deep breathing techniques by asking them to

- Place one hand on the upper chest and the other on the upper abdomen, then breathe. Notice how your breath moves up and down through the diaphragm.
- Take a deep breath through your nose, feeling your stomach rise and fall beneath your hand.

- Next, engage your core as you breathe in through your nose and out through pursed lips.

Over time, practicing these simple healing techniques for sensorimotor connection helps survivors reduce the trauma-driven reactions they have carried since the onset of their traumatic experiences. Addressing maladaptive responses contributes to their inability not only to modulate their emotions but also to relax, alleviating both physical and emotional pain. For example, those with a survival mindset, like Ali, struggle with somatic responses that intensify their involuntary emotional impulses, leading to muscle tightness, pain, insomnia, and self-isolation (Ogden, 2021).

As a result, mindfulness meditation for TICBI is essential for promoting recovery by alleviating these challenges. It allows survivors to ease both physical and emotional pain while learning to anchor themselves in the present, focusing on the here and now as they distance themselves from overwhelming and distressing emotions. Grounding techniques and mindful meditation also reduce stress and boost emotional regulation. Grounding helps connect individuals with their physical reality, while mindfulness meditation cultivates attention to the present, promoting calm and reducing reactivity. Grounding activities or culturally informed mindfulness practices, due to their calming effect, help survivors enhance the connection between their brainstem's sensory and survival responses (bottom-up) and cognitive reasoning (top-down).

Some healing practices used in TICBI include visualization and engaging the senses, such as sight and hearing, to redirect survivors from their distressing thoughts and feelings. Additionally, incorporating mindfulness techniques helps survivors remain in the present. These activities utilize visualization and the five senses to assist survivors in refocusing their attention on the present moment. Calming and reframing techniques discussed in this chapter and Chaps. 8 and 9 aid survivors in this learning and adaptation, reconnecting their senses (body and mind) to the present rather than dwelling on the past.

These activities require guided instructions, allowing survivors to engage with the tool in a safe environment. The relationship between the survivor and the therapist facilitates a co-regulation process when clients struggle to let go of the past and learn to live differently. As van der Kolk (2014) notes, "visiting the past in therapy should be done while people are, biologically speaking, firmly rooted in the present and feeling as calm, safe, and grounded as possible" (p. 70), underscoring the benefits of the therapeutic relationship. Utilizing co-regulation to focus on the present moment is vital for the survivor. When they struggle or encounter obstacles in their lives, reminding them to practice being present through breathing and acknowledging their most valued aspects helps bring them back to the present moment, aiding them in learning how to let go of the past. In TICBI, the approach involves opening survivors' minds to new experiences, sensations, and possibilities they can explore in a safe, supportive environment. Attention to sensations and sensorimotor reactions is crucial during deep breathing to promote recovery and eliminate implicit maladaptive responses that have impacted their well-being.

The healing benefits of TICBI depend on the quality of the therapeutic relationship and the clarity of the introduction to each activity. For instance, psychoeducation clarifies how the client's nervous system constantly provides feedback on their actions and reactions, particularly in maintaining protection mode, which can sometimes lead to self-isolation. Sharing information about the functions of cranial nerves originating from the brainstem and their role in helping to protect against perceived threats through senses such as seeing, hearing, tasting, and smelling can foster an engaging conversation. This knowledge enables survivors to comprehend previously confusing issues and nurtures their curiosity about healing and recovery. The discussion redirects their attention from traumatic memories to the functions and roles of these senses in healing. Some of these functions include the role of cranial nerves in trauma response.

Steven Porges (2017) describes cranial nerves as "a pathway through which most information comes into the brain and leaves the brain" (p. 134). Incorporating this information into the discussion on how to use grounding techniques for recovery is essential. Techniques such as employing the five senses, discussed below, help survivors reconnect and activate their parasympathetic nervous system, allowing them to stay present and reduce the dissociative feelings or overwhelming emotions that disrupt their lives. These grounding techniques can assist in recalibrating the survivor's nervous system and shifting them from a heightened state of arousal to a more regulated and grounded state. The activities also promote change, relaxation, resilience, and overall well-being.

- **Sight:** Encourage survivors to sit comfortably and upright, with their legs touching, while looking around and identifying objects that inspire positive feelings or sensations. Discuss their reactions and sensorimotor responses if any negative emotions arise while observing items (e.g., pictures on the wall or other objects).
- **Sound:** I often ask survivors to bring music they enjoyed before and after the war to examine how the sound influences their mood. Their selections may include prayers and religious songs, which typically provide comfort and foster a spiritual connection that promotes self-empowerment. While listening to their favorite music, I encourage them to focus on their emotions and physical sensations and to share their thoughts.
- **Taste:** If you can access it, offer clients candy, juice, or tea at the beginning of the session. Ask them to describe the flavors they select and engage them in conversation about how they feel at this moment as they savor the candy of refreshments in a relaxing environment.
- **Smell:** In many cultures, people use perfumes, scented lotions, deodorants, creams, and aftershave. Ask your client about the emotions they experienced while applying it. Since this is their favorite scent, focusing on the feelings it evokes during application helps redirect their attention to pleasant experiences and the present moment.
- **Touch:** I encourage survivors to bring an item or object with significant meaning. I share my example to illustrate the value and importance of the chosen item.

Inspire your client to focus on the feel and texture of the object, helping ground them in the present moment.

The aforementioned activities provide survivors with neural integration and synaptic linkages by concentrating their attention on sensory perception and the present moment. This assists them in developing a coordinated and balanced approach to emotions and sensations, effectively deactivating fear and protection-based (bottom-up) responses (Seigel, 2007). Regular practice of these activities encourages long-term healing for survivors (Hebb's Rule), facilitating the ongoing repair of damaged or disconnected brain areas. Consistent application of grounding techniques is an effective practice in TICBI, rooted in Hebb's Rule (1949), which asserts that cells that "fire together, wire together." In this context, the term "integrate" is derived from a Latin word meaning "make whole" (Dana, 2020, p. 166), aptly characterizing TICBI as the process of re-learning, re-igniting, and regulating disconnected and dysregulated emotions. It is a healing journey that empowers survivors to re-establish a new autonomic rhythm, enhancing reconnection with others and rebuilding their lives.

Each grounding or deep breathing technique is based on the human brain's inherent need for interconnectedness, making the activities in TICBI meaningful. The three interconnected regions (from bottom to top: brainstem, limbic areas, and cortex) that were dysregulated during trauma can be restored through these culturally informed and mindful activities. They provide survivors with immediate and rewarding responses, boosting motivation to continue practicing. However, the effectiveness of these practices is enhanced when TICBI-trained therapists prioritize safety, especially when working with individuals holding a survival mindset. Grounding activities require preparation for each session to ensure a supportive and co-regulating presence. This is a crucial aspect of TICBI, as some survivors may be unfamiliar with suggested healing practices, including those designed to engage the sensorimotor function through body movements such as yoga, tai chi, Pilates, and breath work. These activities help survivors increase their sensory awareness as they engage with their five senses (sight, hearing, touch, smell, and taste), helping them to remain in the present moment. In contrast, physical activities from the abovementioned activities connect their bodies, minds, and the environment.

Therefore, grounding techniques involving physical activities help survivors manage their dysregulated emotions by fostering a clear and logical interpretation of pre-activity responses and post-activity feelings. In this process, immediate evaluation and feedback provide survivors with a calming and encouraging effect, releasing their physical and psychological tension. The neural integration that follows the practice of these activities continues to advance their healing process, creating the feeling described by Siegel (2007) as they develop and benefit from what Daniel Siegel calls "the triangle of well-being." In these practices, the triangle of well-being arises from the "coherent mind and empathic relationships" they develop during the practice, connecting the mind, body, and brain (p. 289). These practices are suitable and adjustable for all cultures and religious beliefs, providing a holistic and culturally informed approach to engage clients. In this culturally informed intervention, survivors learn how to manage their physical, mental, and emotional responses.

TICBI facilitates long-term adaptation by changing survivors' approaches to protection and safety, helping them learn healthier responses to manage their emotions and behaviors while fostering self-confidence and emotional balance. Grounding techniques for sensorimotor connection are among the tools used in providing TICBI, and they contribute to long-term healing and adaptation by reconnecting and strengthening the vertical (top to bottom) and horizontal (left to right) neural links disrupted by traumatic experiences. Recovery emerges from this reconnection, enabling survivors to enhance sensorimotor awareness and emotional flexibility, which helps them embrace new experiences, re-establish social connections, and access resources to further their recovery.

3 Sensorimotor Awareness for Recovery

In TICBI, working on sensorimotor awareness helps survivors connect with their body's responses and inherent wisdom for healing. Physical activities like yoga, tai chi, dance, and mindful movement provide survivors with the sensorimotor awareness and emotional release they need for recovery. These activities help individuals become more attuned to their bodies, release pent-up tension, and regain control and safety. Each physical activity demands full presence and attention to the sensations generated by these activities while valuing the interplay of muscle movement, breathing, renewed energy, and enjoying the well-being that follows each session. Practicing these activities to enhance awareness of physical responses begins with recognizing tense areas in the body, redirecting energy toward those areas, and increasing relaxation. Survivors often appreciate the revitalizing effects of physical activity and the energy it provides. With supportive therapeutic interaction, survivors adapt to this new way of living, transforming their lives as they continue incorporating the techniques.

An activity I learned for sensorimotor awareness occurred during a fitness seminar I attended two decades ago. It focused on stress reduction through touch and massaging the forehead (neurovascular point) while concentrating on the tension and stress one is experiencing. Massaging the neurovascular points while holding, massaging, thinking, and breathing assists survivors in suspending their physical and emotional stress. This technique activates the prefrontal cortex by increasing blood flow to the area, simultaneously creating a top-down regulation of the amygdala, which deactivates its stress response and facilitates clearer thinking and a focus on finding solutions to challenges. Some of the guided instructions I used to increase sensory-motor awareness when working with clients include:

- Use your right finger to hold the right side of your forehead and your left finger to hold the left side.

- Gently massage both sides of your forehead with your fingertips to relieve pressure.
- Apply gentle pressure to your forehead while massaging it. Then, take a deep breath and hold it for a count of three. After that, breathe normally for a count of three.
- Consider a recent stressful event. Inhale while holding your breath for a count of three, then exhale and release it from your body, mind, and emotions for a count of three.
- Place your hands on your knees and relax, inhaling and exhaling for five minutes. If you continue to experience physical or emotional stress, repeat the process.

As discussed in Chap. 4, survivors' recovery and well-being depend on reducing their stress and fear responses, regulating the vagal system, enhancing prosocial calming behaviors, and boosting prefrontal activation to re-evaluate their lives. Body awareness increases the connection between tense and disconnected areas in the body while promoting the release of oxytocin (a hormone that plays a significant role in creating social bonds and establishing relationships).

3.1 Emotional-Centeredness for Recovery

In TICBI, emotional centeredness is required to decrease fear responses and physical and emotional tension (created by sympathetic mobilization) and move survivors toward a ventral vagal state that promotes social connection. Working on recovery is a long process in TICBI; therefore, patience is essential for practicing the activities for intervention and overcoming maladaptive survival behaviors. Some of the activities discussed in this chapter may be challenging for some survivors, as they are designed to decenter adaptive responses and facilitate learning new healing processes. When I began to use breathing techniques to overcome the impact of my prolonged trauma, which caused me physical pain in my major joints (i.e., shoulders and hips) and dysregulation, focusing on breathing was initially difficult for me because I had adapted to holding my breath due to my active sympathetic response. I learned to hold my breath and release it to decrease painful physical tension. Similarly, I slowly learned to use body scanning techniques to release tension in those areas through breath. The following steps help introduce survivors to the practice. As they notice the ease and relief of stress and pain in their bodies, this will encourage them to continue practicing.

- If your physical pain allows you to sit up straight, make sure your back touches the chair or the wall.
- Raise your shoulders and keep them straight. Place your hands on your rib cage as you breathe in and out slowly.
- Continue breathing in this way. Count up or down as you breathe: inhale, 1, 2, 3, 4, 5—exhale, 1, 2, 3, 4, 5. This helps maintain the oxygen levels in your brain and prompts your vagus nerve to relax and reconnect (see Siegel, 2021).

- Take slow, deep abdominal breaths to scan your body for tension and exhale to release it. Focus on the quality of your breath and bodily sensations as you release stress.
- Focus on your breathing while quieting the chatter in your mind. Relax your body to alleviate painful muscle and joint tension.

The practice helps survivors enhance their awareness and learn to manage daily stressors generated by overwhelming memories and mental chatter. It also supports them in coping with triggers, such as news events that constantly remind them of their past experiences. I repeat these activities as needed until I achieve a relaxed and emotionally centered state and instruct my clients to do the same. Referring to Hebb's Rules, repetition strengthens neural connections as individuals unlearn past survival reactions and promote healing. These activities have helped clients shift from emotional and behavioral patterns of protection to patterns of connection, socialization, centeredness, an active ventral vagal state, and post-traumatic growth, just as they did for me. Shifting survivors from patterns of protection to connection patterns is a valuable intervention often used in somatic therapy. It involves using various techniques, such as breathing, to regulate the nervous system and enhance survivors' awareness of their body, brain, and mind. Breathing fosters calmness, regulates survivors' dysregulated emotions, and helps them manage their physical and emotional responses, promoting safety and grounding.

3.2 *Breath for Recovery*

The challenge for survivors to practice emotional centeredness lies in the physiological and psychological impacts of trauma, which hinder their ability to regulate their breathing and often lead to anxiety and panic. Therefore, in TICBI, incorporating breathing techniques into the therapeutic practice promotes centeredness and enhances the activation of the ventral vagal system. This increased activation helps survivors relax their bodies and center their troubled or scattered emotions. Because breathing is culturally congruent, it appeals to everyone and requires minimal physical effort. However, it is important to remember that breathing and relaxation can initially be challenging for survivors. As a result, a slow introduction is necessary, followed by a demonstration and simple instructions. This can be practiced while sitting comfortably or on a yoga mat, focusing on the breath and internal or external sensations. In TICBI, breathing activities can be done indoors or outdoors. The following are some helpful guided techniques I have used for indoor practices.

- Sit in a chair with a tight stomach and straight back, or lie flat on the floor. Take a deep breath and concentrate on the internal energy flow within your body.
- Do you feel the external breeze outside your body? If you notice the airflow as you breathe, pay attention to how the air enters your body through your nose and exits through your mouth.

- Can you feel the sensations in your breath and the changes in your body? As you breathe, pay attention to the beautiful images in your mind and the airflow across your body. Are you aware of your body temperature? Do you notice feelings of warmth or cold?
- As you concentrate on your body's temperature, can you recognize areas of tension? Where is the tension located: in your neck, jaw, shoulders, arms, hips, legs, or feet? Breathe through the tension, focusing on those areas to relieve any discomfort. Do you notice any differences?

As your clients practice these techniques, discuss the identified tight and painful areas in their bodies to learn more about releasing tension through breathing, calming emotions, and promoting change and adaptation. The use of simple introductory activities aids their adaptation, transforming physical tightness (somatic muscular trauma), which causes muscle tension and pain or discomfort. This process provides them with a chance to recognize the impacts of trauma and to set aside painful emotional and physical memories through breathing and emotional centeredness. For example, autogenic breathing practices are part of the therapeutic tools used in TICBI to promote relaxation and healing.

3.3 Autogenic Breathing for Sensorimotor Connection and Recovery

As discussed above, survivors of prolonged trauma often carry the tension and stress of their experiences in their bodies. This can lead to muscle and joint pain as they tense specific areas. Autogenic breathing helps individuals focus on particular areas and relax them, promoting recovery. In 1932, a German psychiatrist, Johannes Heinrich Schultz, introduced the autogenic technique to relieve physical pain through deep breathing and emotional centeredness. It is a creative technique used to balance the body and brain and to foster cognitive change by regulating the nervous system and promoting relaxation. It enables individuals to regain control by being in a calm and focused state of mind to pay attention to bodily sensations. In TICBI, social workers use this technique to reduce stress and insomnia, enhance emotional control by quelling mental chatter, improve attention, and normalize blood flow. The guided activities in autogenic breathing include verbal instruction asking survivors to

1. Sit up straight and relax, with eyes open or closed—whatever is comfortable and feels safe.
2. Pay attention to the breath (in and out) and notice how it feels.
3. Pay attention to the body, observing tension and discomfort without judgment.
4. Pay attention to your large muscles and notice the sensations in those areas. Feel the tension or weight in those areas as you breathe in and out to relax them.

The goal of using these verbal commands for autogenic breathing is to help survivors achieve a state of deep relaxation, thereby reducing their trauma-related stress and enabling them to articulate feelings and physical sensations they were previously unable to address or explain. Therefore, it is unsurprising to hear survivors saying at the end of the practice:

- I feel safe and relaxed.
- I can feel my breath going in and out calmly and regularly.
- I feel calm, and my heart rate is regular.
- My major muscles, where I often feel tension, are now relaxed.
- My stomach is no longer upset but feels comfortable.
- My shoulders are now less tense and relaxed.
- My joint pain has subsided.
- My body temperature is now average.

It is important to note that teaching survivors to focus on their body parts while engaging in deep breathing exercises can initially be challenging due to the sensory-motor amnesia (SMA) they have developed for survival. SMA often impacts those experiencing trauma at a young age, affecting their sensory awareness and muscular control. Childhood trauma is known to disrupt sensory development, involving the capacity for sensory processing and emotional regulation.

The difficulties posed by these adaptive and protective sensorimotor responses become evident when practicing the pendulation and titration techniques mentioned earlier. Autogenic breathing helps survivors enhance their top-down brain activation, modifying the connections that lead to muscle tightness and dysregulation through guided interventions.

As discussed above, visualizing and verbalizing sensorimotor responses during various types of breathing exercises and mindfulness practices places survivors in a relaxed ventral vagal state, enabling them to experience the benefits of emotional and physical balance and regulation. Incorporating additional physical activities into the autogenic breathing exercises helps alleviate body–brain tension and strengthen their connection, leading them away from lifelong suffering and a persistent sense of protection (Siegel, 2010, 2021).

These suggested exercises for prolonged trauma are designed to help survivors manage their overactive nervous systems, anxiety, and hypervigilance, which continue to impact their quality of life. For example, autogenic breathing is incorporated into TICBI to enable survivors to regulate their emotions by calming their sympathetic nervous system (the "fight-or-flight" response) and promoting relaxation. In TICBI, guided activities are particularly beneficial during the initial stages of the intervention, as they assist survivors in centering themselves and adapting the activities for extended use. Guided activities such as breathing and physical interventions empower survivors as they recognize their strength and serene state, allowing them to enhance their focus on moment-to-moment feelings and control trigger responses. Consistent breathing and physical activities over time also improve the reconnection of the body, brain, and mind (Hebb's Rule: neurons that fire together, wire together), enabling survivors to alleviate somatic experiences and support their bodily, cognitive, and emotional recovery.

4 Physical Activities for Sensorimotor Connection and Recovery

The connection between the body, brain, and mind illustrates how our feelings and thoughts influence our physical experiences in life. Like reframing, narratives shift the survivor's focus from negative thought patterns to positive ones; the human body can be retrained to relax and relieve the physical tension that causes somatic pain. Engaging in physical activities is part of this retraining, allowing the body to naturally rewire itself, enhancing blood flow, and increasing muscle flexibility and relaxation. Sensorimotor therapy, developed by Dr. Pat Ogden and her colleagues in the 1980s, employs this retraining process through a body-brain-mind-centered approach that emphasizes activities increasing the awareness and tracking of the "five building blocks": sensations, emotions, thoughts, images, and movement, as identified in the client's verbal narratives (Ogden, 2021, p. 133).

Physical activities and mindfulness practices enhance awareness of these building blocks and their impacts on the body, brain, and mind. According to Levine (2015, p. 115), because humans express their emotional experiences beyond verbal communication, it is crucial to observe physical responses such as facial expressions and body gestures (e.g., anger: clenched fists and gritted teeth; fear: tight muscles and shallow breathing) to mitigate their impacts. As mentioned earlier, the repeated manifestation of these sensorimotor responses creates tension that can contribute to physical pain. Survivors often report post-traumatic physical pain affecting their quality of life.

For decades, renowned trauma researchers like Peter Levine (2015), Bessel van der Kolk (2014), and Pat Ogden (2021) have explored how trauma affects sensorimotor responses, leading to physical pain. As a result, physical activities are recommended for TICBI to relieve tension and alleviate pain while enhancing survivors' self-awareness (mediated by the medial prefrontal cortex) and body awareness (facilitated by the insula). These activities provide the benefit that research suggests: "brain areas that convey pain and distress are also responsible for transmitting feelings of joy, pleasure, purpose, and relational connection" (Levine, 2015, p. 115). Therefore, integrating reframing survival narratives, breathing exercises, and physical activities to reduce tension in TICBI is a vital aspect of the change talk/talking cure, minimizing pain and distress while strengthening responses that promote joy, pleasure, purpose, well-being, and actions of post-traumatic growth (PTG).

These activities are particularly beneficial for individuals with a survival mindset who experience muscle tension due to constant hyperarousal or hypervigilance. Physical activities may not be suitable for some survivors and require a gradual introduction with clear instructions to prevent miscommunication and disappointment. Information on how physical activities and mindfulness practices contribute to relieving accumulated body tension and the healing process is encouraging and motivating. Physical activity centers emotions and promotes being in the present moment. It is part of the practices in TICBI that are used for holistic healing and enhancing survivors' well-being. It increases blood flow to disconnected areas and reduces disharmony among the body, brain, and mind—the sources of somatic pain. However, as every trauma intervention requires, starting each activity with simple, slow techniques tailored to each client's abilities helps them recognize their implicit physical and emotional responses.

Studies have shown that psychotherapeutic interventions, particularly those adopting a body-centered approach that emphasizes the body–brain–mind connection, provide both immediate relief and promote long-term adaptive lifestyle changes (Ogden & Fisher, 2015; Siegel, 2007, 2010, 2011; van der Kolk, 2014). To ensure that physical activities are acceptable, accessible, and meaningful, it is crucial to consider the client's background, which includes family, culture, history, and physical capacity. Activities such as breathing exercises, yoga, tai chi, Pilates, and dance are integral to somatic therapy and represent a creative approach that can be incorporated into the practice of TICBI. This approach reduces cognitive, emotional, and physiological symptoms linked to trauma and PTSD while encouraging therapeutic conversations about multiple healing practices (Siegel, 2007; van der Kolk, 2014). These activities can decrease muscle tightness commonly associated with PTSD and improve the neuroendocrine system linked to the production of stress hormones. Clients can use self-regulating activities that independently connect the body and brain as part of take-home activities and long-term interventions.

Furthermore, incorporating physical activities in TICBI enhances survivors' cognitive capacity and aids in monitoring and assessing their sensorimotor responses (top-down). Simultaneously, breathing, grounding, and mindfulness meditation help regulate emotions (bottom-up), recalibrating the autonomic nervous system (van der Kolk, 2014). This process releases the chronic physical tension from constant fear, hyperarousal, or hypoarousal (Kurtz, 1990). The body, brain, and mind-based intervention in TICBI fosters emotional balance and grants access to the neocortex, enabling survivors to become curious about change and open to exploration and growth. In TICBI, all proposed activities aim to encourage survivors to reconnect the disjointed parts of their body, brain, and mind while enhancing their health and well-being. However, when selecting activities for TICBI, there are certain precautions to consider:

- The client's physical and emotional limitations and readiness.
- The client's prior involvement in physical activities (e.g., cultural dance or body movements, religious chants) can make the change-based activities more acceptable. When introducing new ideas and healing practices, it is crucial to factor in your client's resilience, strength, drive for survival, and emotional flexibility.

These inner strengths enhance survivors' motivation and commitment to sustain the practice and change their lives.

Definitions

Affect (emotion) regulation: The ability to modulate one's emotional state to adapt and meet environmental demands.

Autogenic breathing: A relaxation technique using deep breathing. It begins with exhaling completely, taking a small breath through the nose, holding it for three seconds, and exhaling again. This process is repeated until relaxation is felt.

Bottom-up information flow: Incoming sensory data (from the bottom part of the brain) through the brainstem, moving upward to the cortex.

Centeredness: Being mentally and emotionally balanced and focused in mindfulness.

Co-regulation: The nervous systems of two individuals interacting to show support and emotional connection through mutual regulation of their physiological state.

Epigenetics: The study of how behaviors and environments affect genes. The DNA sequence does not change, but how the body reads a DNA sequence does. Unlike genetic changes, these changes are reversible through intervention.

Hippocampus: The brain area that is responsible for learning and storing memory.

Interoception: The ability to identify, access, understand, and respond to patterns of internal sensations. This provides a distinct advantage in overcoming challenges and enhancing adaptation.

Medial prefrontal cortex (mPFC): The part of the brain mediating decision-making and memory consolidation.

Meditation is a practice-based technique for increasing awareness, detaching oneself from "brain chatter and negative thinking," improving mental clarity and emotional calmness, and being homeostatic.

Neocortex: The six-layered part of the cortex involved in sensory perception, cognition, generation of motor commands, spatial reasoning, and language.

Neuroendocrine system: A network of cells that produce hormones when signaled by the nervous system.

Sensorimotor: The integration of sensory information and motor responses.

Somatic therapy: A therapeutic method promoting mental health focused on the physical expression of emotions.

Somatoform: Experiencing physical symptoms causing discomfort and pain that cannot be explained.

Top-down information flow: The brain's reflective capabilities to act based on external information to influence the present state of mind, memories, and emotions.

Discussion Questions

1. What are some of the techniques mentioned in this chapter as part of the TICBI recovery process?
2. What are some of the breathing exercises used to create a lasting change?
3. Based on clients you have worked with in the past or are currently working with, plan a short intervention using the skills you learned about in this chapter.

4. Reflect on the intervention tools discussed in this chapter and their applicability to TICBI. Will you be able to use them in your future practice? If not, what will you modify to support better the survivors you are or will be working with?

Reflective Exercises

1. If you have been involved in community or clinical interventions, list the trauma-informed activities you have used.
2. Was there a particular activity you used that worked well for survivors of prolonged war and trauma? Think about those activities and compare and contrast them with what you have learned from this chapter.
3. Did you discover a difference between what you used in the past and what you learned from this chapter? Is there an activity that stands out to you?
4. Practice the suggested breathing and body movement activities, and list the benefits and challenges you experience as you practice.
5. How will you adjust the activities described in the chapter based on the survivor's interests, physical limitations, and fears as a first-time service user, and how will you overcome these?

Competing Interests The author has no conflicts of interest to declare that are relevant to the content of this chapter.

References

Applegate, J. S., & Shapiro, J. R. (2005). *Neurobiology for clinical social work*. W. W. Norton & Company.

Cozolino, L. J. (2010). *The neuroscience of psychotherapy: Healing the social brain* (2nd ed.). Norton & Company.

Dana, D. (2020). *Polyvagal exercises for therapists and clients: 50 Client-Centered Practices*. W. W. Norton & Co.

Farmer, R. L. (2008). *Neuroscience and social work practice*. SAGE Publications.

Hanna, T. (1986). What is somatics? *Somatics: Magazine-Journal of the Bodily Arts and Sciences, 5*(4), 5–8.

Hanna, T. (1988). *Somatics: Reawakening the mind's control of movement, flexibility, and health*. Da Capo Press.

Hanna, T. (1990). Clinical Somatic education: A new discipline in the field of health care. *Somatics: Magazine-Journal of the Bodily Arts and Sciences, 8*(1), 4–10.

Hebb, D. O. (1949). *The organization of behavior: A Neuropsychological Theory*. Wiley.

Herman, J. L. (1997). *Trauma and recovery*. Basic Books.

Kurtz, R. (1990). *Body-centered psychotherapy: The Hakomi Method*. Life Rhythm.

Levine, P. (2015). *Trauma and memory: Brain and body in a search for the living past: A practical guide for understanding and working with traumatic memory*. North Atlantic Books.

Ogden, P., Minton, K., & Pain, C. (2006). *Trauma and the Body: A Sensorimotor Approach to Psychotherapy (Norton Series on Interpersonal Neurobiology)*. (1st ed.). W. W. Norton & Company.

Ogden, P. (2021). *The pocket guide to sensorimotor psychotherapy in context (Norton Series on Interpersonal Neurobiology)*. W. W. Norton & Company.

Ogden, P., & Fisher, J. (2015). *Sensorimotor psychotherapy: Interventions for trauma and attachment (Csm)*. W. W. Norton & Company.
Porges, S. W. (2017). *The pocket guide to the polyvagal Theory: The Transformative Power of Feeling Safe*. W.W. Norton.
Porges, S. W., & Dana, D. (2018). *Clinical applications of the polyvagal theory: The emergence of Polyvagal-Informed therapies (Norton series on interpersonal neurobiology)*. W. W. Norton & Company.
Rogers, C. R. (1951). *Client-centered therapy: Its current practice, implications and theory*. Houghton Mifflin.
Selye, H. (1976). Stress and physical activity. *McGill Journal of Education, 11*(1), 3.
Shapiro, E., & Shapiro, D. (2019). *The unexpected power of mindfulness and meditation*. Ixia Press.
Siegel, D. J. (2007). *The mindful brain: Reflection and attunement in the cultivation of Well-Being*. Norton & Company.
Siegel, D. J. (2010). *The mindful therapist: A clinician's guide to mindsight and neural integration (Norton Series on Interpersonal Neurobiology)*. W. W. Norton & Company.
Siegel, D. J. (2011). *Mindsight: The new science of personal transformation*. Bantam.
Siegel, D. J. (2021). *Becoming aware: A 21-day mindfulness program for reducing anxiety and cultivating calm*. Penguin Random House.
Van Der Kolk, B. A. (2014). *The body keeps the score: Brain, mind, and body in the healing of trauma*. Penguin Publishing Group.

Index

N. Bokore, *Trauma-Informed, Culturally Based Intervention*,
https://doi.org/10.1007/978-3-032-06835-4

MIX
Papier aus verantwortungsvollen Quellen
Paper from responsible sources
FSC® C105338

If you have any concerns about our products,
you can contact us on
ProductSafety@springernature.com

In case Publisher is established outside the EU,
the EU authorized representative is:
Springer Nature Customer Service Center GmbH
Europaplatz 3, 69115 Heidelberg, Germany

Printed by Libri Plureos GmbH
in Hamburg, Germany